43 iwe 300
ebf 548
2. Exemplar

Ausgeschieden im Jahr 2025

The g_m/I_D Methodology,
A Sizing Tool for Low-voltage
Analog CMOS Circuits

ANALOG CIRCUITS AND SIGNAL PROCESSING SERIES

Consulting Editor: **Mohammed Ismail.** *Ohio State University*

For other titles published in this series, go to
www.springer.com/series/7381

The g_m/I_D Methodology, A Sizing Tool for Low-voltage Analog CMOS Circuits

The Semi-empirical and Compact Model Approaches

By

Paul G.A. Jespers
Université Catholique de Louvain
Louvain-la-Neuve, Belgium

Prof. Paul G.A. Jespers
Université Catholique de Louvain
Louvain-la-Neuve
Belgium
Paul.Jespers@uclouvain.be

Additional material to this book can be downloaded from http://extra.springer.com.

ISBN 978-0-387-47100-6   e-ISBN 978-0-387-47101-3
DOI 10.1007/978-0-387-47101-3
Springer Dordrecht Heidelberg London New York

Library of Congress Control Number: 2009940107

© Springer Science+Business Media, LLC 2010
All rights reserved. This work may not be translated or copied in whole or in part without the written permission of the publisher (Springer Science+Business Media, LLC, 233 Spring Street, New York, NY 10013, USA), except for brief excerpts in connection with reviews or scholarly analysis. Use in connection with any form of information storage and retrieval, electronic adaptation, computer software, or by similar or dissimilar methodology now known or hereafter developed is forbidden.
The use in this publication of trade names, trademarks, service marks, and similar terms, even if they are not identified as such, is not to be taken as an expression of opinion as to whether or not they are subject to proprietary rights.

Printed on acid-free paper

Springer is part of Springer Science+Business Media (www.springer.com)

to Denise
and
to my parents
Oscar Jespers
and Mia Carpentier

Foreword

IC designers appraise currently transistors sizes while having to fulfill simultaneously a large number of objectives like a prescribed gain-bandwidth product, minimal power consumption, minimal area, low-voltage design, dynamic range, non-linear distortion, etc. Making appropriate decisions is not always obvious. How to meet gain-bandwidth specifications while minimizing power consumption of an Op. Amp without area penalty? Should moderate inversion be preferred to strong inversion? Is sizing an art or a mixture of design experience and repeated simulations? Or is it a constrained multivariate optimization problem? Optimization algorithms are attractive without doubt but they require translating not always well-defined concepts into mathematical expressions. The interactions amid semiconductor physics and systems are not always easy to implement.

The objective of the book is to devise a methodology enabling to fix currents and transistors widths of CMOS analog circuits so as to meet specifications such as gain-bandwidth while optimizing attributes like low power and small area. A special attention is given to low-voltage circuits. The sizing method takes advantage of the g_m/I_D ratio and makes use of either 'semi-empirical' data or compact models. The 'semi-empirical' approach utilizes large look-up tables derived from physical measurements carried out on real transistors or advanced models. The compact model approach offers the possibility to make use of analytic expressions. Unfortunately when it comes to real transistors, especially sub-micron devices, this isn't true anymore. Other means are necessary to keep track of high order effects without the risk to loose the inherent simplicity of compact models. Bias dependent instead of constant parameters offer the possibility to extend the validity of a model like the E.K.V. model.

In the first chapter, the Intrinsic Gain Stage, is sized making use of the classical strong and weak inversion large signal models of MOS transistors. This leaves open the moderate inversion region, a region that offers the best compromises generally as far as power consumption and sizes. To be able to size circuits in moderate inversion, we need a reliable large signal MOS model. The Charge Sheet Model that is considered in Chapter 2 is an invaluable tool for understanding the mechanisms governing current in MOS transistors, but it is not fitted for real transistors for it relies on the gradual channel approximation and makes use of mathematical expressions that are too complicated. The MATLAB tools that are available under 'extras.springer.com' overcome the mathematical aspects and offer the possibility to perform 'ideal experiments'. Some of the abstract aspects of the Charge Sheet Model moreover are bridged in Chapter 3 by the introduction of a graphical representation of the drain current that combines physical aspects and practical circuits.

The E.K.V. basic model discussed in Chapter 4, offers clearly more flexibility. It is an approximation of the Charge Sheet Model and a forerunner of what is viewed nowadays as compact Surface Potential Models. The model paves the way towards analytical expressions not only for the drain current but also for the terminal voltages whatsoever the mode of operation of the transistor, whether saturated or not. Unfortunately, the simple E.K.V. model is a gradual channel model like the Charge Sheet Model, unfit thus for real transistors, in particular short channel device.

The fact that drain currents predicted by the E.K.V. compact model look so similar to real drain currents opens the question whether the model could not be extended to real devices. In Chapter 5, we show that currents very close to real drain currents can be predicted when the parameters of the E.K.V. model vary with bias, even with 100 nm devices. The explanation may be the quasi-one-dimensional nature of the channel opposed to the two-dimensional space charge below the inversion layer. As a result, gradual channel conditions prevail in the inversion layer any longer than in the space charge when the gate length is shrinking. An algorithm is proposed to acquire the model parameters.

The Intrinsic Gain Stage is reconsidered in Chapter 6 in the light of the variable parameters compact model. Currents and transistor width obtained by means of the compact model reproduce very closely the values obtained by means of the 'semi-empirical' method. A series of examples considering a low-frequency and a one GHz gain-bandwidth product I.G.S. are described.

The remaining Chapters 7 and 8 extend the method respectively to the common-gate stage and to the basic Miller Op. Amp. The latter illustrates how to meet both, specifications and attributes. Specifications concern the gain-bandwidth product and phase margin, attributes low power and area. These determine optimal regions in the 2D sizing space defined by the first and second stages of the Miller Op. Amp. A MATLAB file compares design strategies.

I want to express my gratitude to Piet Wambacq for the opportunity he gave me to check the validity of the variable parameter E.K.V. model on a 90 nm technology developed by IMEC. I am also very thankful Prof. Gilbert Declerck, former President CEO and Ludo Deferm, executive vice-president of IMEC, who gave me permission to publish the results and the data listed under the 'extras.springer.com'.

My sincere thanks go to Prof. Fernando Silveira who published in 1996 the first paper illustrating the potential of the g_m/I_D methodology. I want to thank him as well as Prof. A. Vladimirescu for the very detailed comments and suggestions they made of the first chapters. I also want to associate Prof. D. Flandre to my thanks owing to our long-term collaboration at the Microelectronics lab of the Université Catholique de Louvain.

Though the specific current put to use in the book is the one defined in the E.K.V. model, I owe much to two research groups. I am indebted to Prof. Eric Vittoz for the

E.K.V. model, and to Prof. Carlos Galup-Montoro and Marcio C. Schneider for the A.C.M. model. I thank the supporters of the two models for motivating discussions and in particular the opportunity Prof. Montoro and Schneider gave me to visit them at the Federal University of Santa Catarina, Brasil.

Tervuren, July 2009   *P. Jespers*

Contents

1 **Sizing the Intrinsic Gain Stage** 1
   1.1   The Intrinsic Gain Stage 1
   1.2   The Intrinsic Gain Stage Frequency Response 1
   1.3   Sizing the Intrinsic Gain Stage 3
        1.3.1   Sizing the I.G.S. with the Quadratic Model 4
        1.3.2   Sizing the I.G.S. by Means of the Weak
                Inversion Model 4
        1.3.3   Sizing the I.G.S. in the Moderate Inversion Region 5
   1.4   The g_m/I_D Sizing Methodology 7
   1.5   Conclusions 8

2 **The Charge Sheet Model Revisited** 11
   2.1   Why the Charge Sheet Model? 11
   2.2   The Generic Drain Current Equation 11
   2.3   The Charge Sheet Model *Drain Current Equation* 13
   2.4   Common Source Characteristics 15
        2.4.1   The $I_D(V_D)$ Characteristics 15
        2.4.2   The $I_D(V_G)$ Characteristic of the Saturated Transistor 17
        2.4.3   Drift and Diffusion Contributions to the Drain Current 18
   2.5   Weak Inversion Approximation of the Charge Sheet Model 18
   2.6   The g_m/I_D Ratio in the Common Source Configuration 20
   2.7   Common Gate Characteristics of the Saturated Transistor 23
   2.8   A Few Concluding Remarks Concerning the C.S.M. 24

3 **Graphical Interpretation of the Charge Sheet Model** 25
   3.1   A Graphical Representation of I_D 25
   3.2   More on the V_T Curve 28
   3.3   Two Approximate Representations of V_T 29
        3.3.1   The 'Linear' Surface Potential Approximation 29
        3.3.2   The 'Linear' Threshold Voltage V_T Approximation 31
   3.4   A Few Examples Illustrating the Use of the Graphical Construction 32
        3.4.1   The MOS Diode 32

	3.4.2	The MOS Source Follower	32
	3.4.3	The CMOS Inverter	33
	3.4.4	Small Signal Transconductances	34
	3.4.5	CMOS Transmission Gates	35
	3.4.6	How to Implement Quasi-linear Resistors with MOS Transistors	36
	3.4.7	Source-Bootstrapping	37
3.5	A Closer Look to the Pinch-Off Region		38
3.6	Conclusion		39

4 Compact Modeling ... 41

4.1	The Basic Compact Model		41
4.2	The E.K.V. Model		42
	4.2.1	The $V_T(V)$ Characteristic	42
	4.2.2	The Drain Current	45
	4.2.3	The Equations of the Basic E.K.V. Model	46
	4.2.4	Graphical Interpretation of the E.K.V. Model	47
4.3	The Common Source Characteristics $I_D(V_G)$		48
4.4	Strong and Weak Inversion Asymptotic Approximations Derived from the Compact Model		50
4.5	Checking the Compact Model Against the C.S.M.		50
	4.5.1	The Acquisition Algorithm (MATLAB Identif3.m)	50
	4.5.2	Verification	52
4.6	Evaluation of g_m/I_D		54
4.7	Sizing the Intrinsic Gain Stage by Means of the E.K.V. Model		55
4.8	The Common-Gate g_{ms}/I_D Ratio		57
4.9	An Earlier Compact Model		58
4.10	Modeling Mobility Degradation		59
	4.10.1	The Impact of Mobility Degradation on the Drain Current	59
	4.10.2	The Impact of Mobility Degradation on the g_m/I_D Ratio	64
	4.10.3	Sizing the Intrinsic Gain Stage in the Presence of Mobility Degradation	65
4.11	Conclusion		66

5 The Real Transistor ... 67

5.1	Short Channel Effects		67
5.2	Checking the Validity of the Compact Model when its Parameters vary with the Source and Drain Voltages		69
	5.2.1	E.K.V Parameter Identification (MATLAB IdentifDemo.m)	70
	5.2.2	How to Introduce Mobility Degradation?	73
	5.2.3	Drain Current Reconstruction	75

	5.3	Compact Model Parameters Versus Bias and Gate Length	76
		5.3.1 The Influence of the Gate Length on the Model Parameters	76
		5.3.2 The Influence of Bias Conditions on the Parameters	78
	5.4	Reconstructing $I_D(V_{DS})$ Characteristic	82
	5.5	Evaluation of g_x/I_D Ratios	84
		5.5.1 The g_m/I_D Ratio	85
		5.5.2 The g_d/I_D Ratio	88
	5.6	Conclusions	91
6	**The Real Intrinsic Gain Stage**		**93**
	6.1	The Dependence on Bias Conditions of the g_m/I_D and g_d/I_D Ratios (MATLAB fig061.m)	93
	6.2	Sizing the I.G.S with 'Semi-empirical' Data	94
		6.2.1 Sizing the I.G.S Loaded by a Constant Total Capacitance	95
		6.2.2 Introduction of Extrinsic Capacitances	99
		6.2.3 Sizing the I.G.S Loaded by a Constant Load Capacitance	103
	6.3	Model Driven Sizing of the I.G.S.	104
		6.3.1 Sizing W and ID (MATLAB fig612.m)	104
		6.3.2 Evaluation of the Intrinsic Gain (MATLAB fig613.m)	106
		6.3.3 An Alternative Method to Evaluate the Gain (MATLAB fig615.m)	107
		6.3.4 A Simplified Sizing Procedure	110
	6.4	Slew-Rate Considerations	111
	6.5	Conclusions	112
7	**The Common-Gate Configuration**		**113**
	7.1	Drain Current Versus Source-to-Substrate Voltage (Matlab fig071.m)	113
	7.2	The Cascoded Intrinsic Gain Stage	115
		7.2.1 Sizing the Cascode (Matlab fig074.m)	115
		7.2.2 Gain Evaluation of the Cascode (MATLAB fig075.m)	117
		7.2.3 The Poles of the Cascode Circuit (MATLAB fig075.m)	118
8	**Sizing the Miller Op. Amp.**		**121**
	8.1	Introductory Considerations	121
	8.2	The Miller Op. Amp.	121
		8.2.1 Analysis of the Miller Operational Amplifier	122
		8.2.2 Pole Splitting	123
		8.2.3 The Impact of the Current Mirror	126
		8.2.4 Poles and Zeros	127

	8.3	Sizing the Miller Operational Amplifier (MATLAB OpAmp.m) 129	
		8.3.1 Sizing a Low-voltage Miller Op. Amp. 130	
		8.3.2 Sizing a High-Frequency Low-Power Miller Op. Amp. 140	
	8.4	Conclusion 142	

Annex 1 How to Utilize the Data available under 'extras.springer.com'...143
- A1.1 Global Variables 143
- A1.2 An Example Making Use of the 'Semi-empirical' Data: The Evaluation of Drain Currents and g_m/I_D Ratio Matrices (MATLAB A12.m)...144
- A1.3 An Example Making Use of the E.K.V Global Variables: The Elaboration of an ID(VGS) Characteristic (Matlab A13.m)...146

Annex 2 The 'MATLAB' Toolbox..149
- A2.1 Charge Sheet Model Files 149
  - A2.1.1 The **pMat(T,N,tox)** Function 149
  - A2.1.2 The **surfpot(p,V,VG)** Function 150
  - A2.1.3 The **IDsh(p,VS,VD,VG)** Function 151
- A2.2 Compact Model Files...151
  - A2.2.1 The **Identif 3(Nb,tox,VFB,T)** Function 151
  - A2.2.2 The **invq(z)** Function..152
  - A2.2.3 The **ComS(VGS,VDS,VS,lg)** Function 152
- A2.3 Other Functions...152
  - A2.3.1 The **jctCap(L,W,R,V)** Function............................152
  - A2.3.2 The **Gss(x,H)** Function 153

Annex 3 Temperature and Mismatch, from C.S.M. to E.K.V. 155
- A3.1 The Influence of the Temperature on the Drain Current (MATLAB A31.m) 155
- A3.2 The Influence of the Temperature on gm/ID (Matlab A32.m)........156
- A3.3 Temperature Dependence of E.K.V Parameters (MATLAB A33.m) 158
- A3.4 The Impact of Technological Mismatches on the Drain Current (Matlab A34.m)..159
- A3.5 Mismatch and E.K.V Parameters (MATLAB A35.m) 161

Annex 4 E.K.V. Intrinsic Capacitance Model 163

Bibliography 167

Index..169

Notations

A, A_{DC}, A_{AC}	voltage gain, DC and AC voltage gain
A.C.M.	Advanced Compact Model
C.L.M.	Channel Length Modulation
C.S.M.	Charge Sheet Model
C	capacitor value
C'_{ox}	gate oxide capacitance per unit area
C_{GB}	gate-to-substrate capacitance
C_{GD}	gate-to-drain capacitance
C_{GS}	gate-to-source capacitance
C_J	junction capacitance
C_{Jsw}	peripheral side-wall junction capacitance
C_{Jswg}	gate side–wall junction capacitance
C_m	Miller capacitance
CMOS	Complementary MOS
D	diffusion constant
D.I.B.L.	Drain Induced Barrier Lowering
E.K.V.	Enz, Krumenacher and Vittoz compact model
G.V.O.	Gate Voltage Overdrive voltage
g_d	output conductance
g_m	gate transconductance
g_{mb}	bulk transconductance
g_{ms}	source transconductance
i, i_F, i_R	normalized drain current, forward and reverse i
I.G.S.	Intrinsic Gain Stage
I_D	DC drain current
I_{Du}	unary DC drain current (W = L)
I_S	specific current
I_{Su}	unary specific current (W = L)
I_{Suo}	weak inversion unary specific current
L	gate length
N	impurity concentration
n	slope factor

PolyN, PolyP	mobility degradation polyn. of N- and P-channel transistors
q, q_F, q_R	normalized mobile charge density, forward and reverse q
q_S, q_D	normalized mobile charge density at the source and drain
Q'_B	bulk charge density
$Q'_i,$	mobile charge density
$Q'_t,$	total charge density
R.H.P.	Right Half Plane zero
S_{VTo}	threshold voltage sensitivity factor with respect to V_{DS}
ThN, ThP	mobility degradation function of N- and P-channel transistors
U_T	thermal voltage kT/q
V, I, v, i	large and small signal voltage or current
V_A	*Early voltage*
V_S, V_G, V_D	source, gate and drain voltage with respect to substrate
V_{GS}, V_{DS}	gate and drain voltage with respect to the source
V_P, V_{PS}	pinch-off voltage with respect to the substrate or the source
v_{sat}	saturation velocity of mobile carriers
V_T	threshold voltage with respect to the substrate
V_{To}	threshold voltage with respect to the source
W	gate width
W.I, M.I, S.I	weak, moderate and strong inversion
β	$\mu C'_{ox}$ *W/L of MOS transistor*
γ	gamma of SPICE program
μ	mobility
μ_o	low-field mobility
ψ_S	surface potential
ω	angular frequency ($2\pi f$)
ω_c	angular cut-off frequency ($2\pi f_c$)
ω_T	angular transition frequency ($2\pi f_T$)

Chapter 1
Sizing the Intrinsic Gain Stage

1.1 The Intrinsic Gain Stage

Sizing methods assessing drain currents and gate widths of a simple circuit are reviewed in this chapter. The circuit, shown in Fig. 1.1, consists of a saturated common source transistor loaded by a capacitor. A constant current source is feeding the drain. The circuit is called currently the 'Intrinsic Gain Stage' (I.G.S.), the name 'intrinsic' underlining the fact that few parts aside the transistor control the performances of the circuit.

Our objective is to find gate widths and drain currents enabling to achieve a prescribed gain-bandwidth product ω_T. We therefore consider the small signal equivalent circuit shown in Fig. 1.2. The input is an open circuit while the output consists of a dependent current source $g_m v_{in}$ (where g_m represents the transconductance of Q) in parallel with the output conductance g_d and the capacitor C.

1.2 The Intrinsic Gain Stage Frequency Response

We divide the I.G.S in high and low frequency sub-circuits to evaluate its frequency response. At high frequencies, all the current delivered by the current source flows through the capacitor for C behaves like a short with respect to the output resistance. Hence:

$$g_m v_{in} = -j\omega \, C v_{out} \tag{1.1}$$

The AC gain is given consequently by:

$$A_{AC} = -\frac{\left(\frac{g_m}{C}\right)}{j\omega} \tag{1.2}$$

At low frequencies, the opposite takes place. The capacitor C is practically an open circuit so that the current flows through the output conductance g_d. Hence:

$$g_m v_{in} = -g_d \, v_{out} \tag{1.3}$$

Fig. 1.1 The 'Intrinsic Gain Stage'

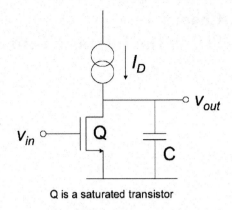

Q is a saturated transistor

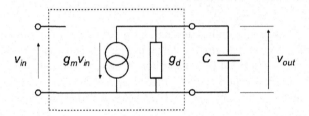

Fig. 1.2 The equivalent small signal circuit of the 'Intrinsic Gain Stage'

The DC gain is given by:

$$A_{DC} = -\frac{g_m}{g_d} \tag{1.4}$$

To get closer to real world transistors, we are going to take into consideration the dependence on bias conditions of the output conductance g_d. Generally, the impact of the current on g_d is acknowledged by replacing the output conductance by the ratio of the drain current over the so-called Early voltage V_A. The Early voltage is supposed to be constant, which implies that all $I_D(V_D)$ characteristics cross the horizontal axis at a unique point once extrapolated. While more or less correct in weak inversion,[1] this is a rather crude approximation in strong inversion, particularly with sub-micron transistors. Our goal being presently to lay down the grass roots of sizing, we are going to assume nevertheless that V_A is constant and postpone more advanced representations to later chapters. Equation 1.4 may be rewritten then as follows:

$$A_{DC} = -\frac{g_m}{I_D} V_A \tag{1.5}$$

---

[1] Weak inversion occurs when MOS transistors are biased with gate voltages lower than the threshold voltage resulting in an exponential relationship between drain current and gate voltage (Vittoz 1977). Strong inversion designates the region where the classic quadratic current to voltage relationship holds true. The transition from weak to strong inversion is currently referred to as the moderate inversion region. This region plays a key role in CMOS analog circuits.

1.3 Sizing the Intrinsic Gain Stage

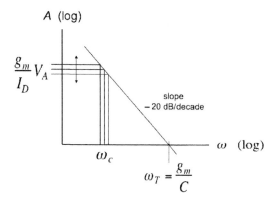

Fig. 1.3 The frequency response of the Intrinsic Gain Stage

Figure 1.3 shows the frequency response of the Intrinsic Gain Stage according to Eq. 1.5 for the low frequency part and according to Eq. 1.2 for the high frequency part.

The point where the two asymptotes cross each other ω_c is called currently the *cut-off angular frequency* and the point ω_T where the high frequency response crosses the horizontal axis (the 0 dB gain point) the *transition angular frequency*:

$$\omega_T = \frac{g_m}{C} \quad (1.6)$$

The name *gain-bandwidth product* is given also to the transition angular frequency for ω_T is equal to ω_c times the gain since the Intrinsic Gain Stage is a true first order system. ω_T is a more significant landmark than ω_C for it characterizes the high frequency behavior of the I.G.S. without the need to know highly unpredictable Early voltages.

1.3 Sizing the Intrinsic Gain Stage

How can one fix the drain current and aspect ratio W/L of the I.G.S so as to achieve a desired *transition frequency* f_T? As far as the transconductance, there is no choice for Eq. 1.6 fixes already g_m:

$$g_m = \omega_T C = 2\pi f_T C \quad (1.7)$$

The problem boils down consequently to find means to connect the drain current I_D and the W/L ratio to the transconductance g_m. A large signal model of the transistor is needed therefore. The first that comes up of course is the classical quadratic MOS model.

1.3.1 Sizing the I.G.S. with the Quadratic Model

The drain current of saturated MOS transistors is given by the well-known quadratic expression:

$$I_D = \beta \frac{(V_G - V_{th})^2}{2n} \tag{1.8}$$

V_{th} being the threshold voltage, while

$$\beta = \mu C'_{ox} \frac{W}{L} \tag{1.9}$$

where

μ	is the mobility of the mobile carries of the channel
C'_{ox}	the gate oxide capacitance per unit-area (the ' meaning capacitance per unit-area)
W and L	respectively the gate width and length
n	the slope factor generally comprised between 1.2 and 1.5

The derivative of I_D with respect to V_G yields the transconductance g_m:

$$g_m = \frac{\partial I_D}{\partial V_G} = \beta \frac{V_G - V_{th}}{n} = \sqrt{\frac{2\beta I_D}{n}} \tag{1.10}$$

W/L and I_D are connected thus to the gain-bandwidth product through g_m. Combining Eqs. 1.9 and 1.10, one has:

$$\frac{W}{L} = \frac{n g_m^2}{2\mu C'_{ox}} \cdot \frac{1}{I_D} \tag{1.11}$$

Instead of a single I_D and W/L, many doublets achieve the desired gain-bandwidth product. We can put forward thus additional objectives, like a large DC gain. Since according to Eq. 1.5 the gain varies like the reciprocal of the drain current, the smaller the drain current, the larger the gain. Not only the gain increases, but the power consumption lessens in the same time. Something is wrong however for zero drain current is supposed to entail infinite gain! In fact, as the current is getting smaller, the transistor enters successively in moderate and weak inversion. The quadratic model does not represent the drain current anymore. Another model is required.

1.3.2 Sizing the I.G.S. by Means of the Weak Inversion Model

In weak inversion, the drain current can be represented by means of an exponential expression (Vittoz 1977):

1.3 Sizing the Intrinsic Gain Stage

$$I_D = I_o \exp\left(\frac{V_G}{nU_T}\right) \qquad (1.12)$$

The transconductance is given then by:

$$g_m = \frac{I_D}{nU_T} \qquad (1.13)$$

where U_T stands for kT/q and k for the Boltzmann constant. To attain the desired ω_T, the drain current must be equal to:

$$I_{D\,\min} = nU_T\, g_m \qquad (1.14)$$

This is a very different result from what we got in strong inversion. The drain current alone fixes the gain-bandwidth product while the aspect ratio has no influence at all. The outcome recalls bipolar transistors for their transition frequency also depends on the collector current only and not on the emitter size (as long as strong injection does not take place of course). MOS transistors in weak inversion and bipolar transistors share indeed a common feature: their currents are mainly diffusion currents.

1.3.3 Sizing the I.G.S. in the Moderate Inversion Region

Sizing in moderate inversion requires a better model. A good candidate is the compact model introduced in Chapter 4, which leads to the expression below demonstrated in Section 4.7:

$$\frac{W}{L} = \frac{n g_m^2}{2\mu C'_{ox}} \frac{1}{I_D - I_{D\,\min}} \qquad (1.15)$$

The expression is valid in all modes of operation, from strong to weak inversion. Suppose we want to design an Intrinsic Gain Stage loaded by a 1 pF capacitor targeting a transition frequency of 100 MHz. The transistor's $\mu C'_{ox}$ and slope factor n are respectively equal to 4×10^{-4} A.V^{-2} and 1.2. Figure 1.4 displays the aspect ratios versus drain current achieving the desired gain-bandwidth product. The result is compared to the strong and weak inversion approximations considered earlier. Below I_{Dmin}, nearly 20 μA, it is impossible to achieve the desired gain-bandwidth product. Above, W/L's conform to a hyperbolic curve whose asymptotes coincide with the strong and weak inversion approximations. But in moderate inversion, large differences are clearly visible with respect to the strong and weak inversion approximations.

The same figure displays also the AC gain predicted by Eq. 1.5 considering an Early voltage of 10 V. Since the gain varies like the reciprocal of I_D, small drain currents mean large gains. When the drain current reaches the minimum given

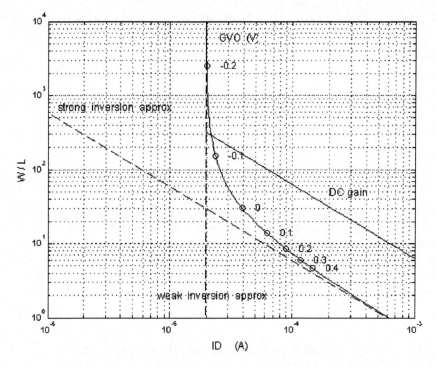

Fig. 1.4 Aspect ratio W/L and AC gain versus the drain current, of an (ideal) Intrinsic Gain Stage aiming at a transition frequency of 100 MHz with a load capacitance of 1 pF. The Early voltage is assumed to be constant and equal to 10. Circles display the difference between gate and threshold voltages, the so-called *Gate Voltage Overdrive* (GVO) (MATLAB fig014.m)

by Eq. 1.14, the gain is largest and equal to the expression below obtained after combining Eqs. 1.5 and 1.13:

$$A_{AC\,\text{max}} = -\frac{V_A}{nU_T} \tag{1.16}$$

Since the thermal voltage U_T at room temperature is only 26 mV, very large gains can be obtained depending on V_A. This once again stresses the commonality shared by bipolar transistors and MOS transistors in weak inversion. The only difference is the n factor. With bipolar transistors, the slope factor is equal to one.

Moderate inversion offers interesting compromises. Currents are smaller than in strong inversion while the W/L ratios are more acceptable than in weak inversion. Gains moreover are only slightly lesser than in weak inversion. Moderate inversion however brings about some drawbacks also. The larger widths that are needed entail more parasitic capacitances than in strong inversion. These require enhancing transconductances, thus also the drain currents, one of the reasons designers kept away from moderate inversion for a long time until the advent of short channel devices.

Where lies the boundary between moderate and strong inversion? To answer the question, consider the $(V_G - V_{th})$ difference, called also the *Gate Voltage Overdrive* (G.V.O.). According to (Lak 1994) strong inversion takes place as soon as the gate voltage overdrive (marked by circles in Fig. 1.4) exceeds 0.2 V. When this happens, moderate inversion W/L's coincide practically with the strong inversion asymptote. Where does weak inversion start? A clear limit is harder to trace, but the fast increase of W/L once the current approaches I_{Dmin} is clearly a sign that weak inversion is taking place. More rigorous limits will be proposed in Chapter 4 when the compact model is introduced.

1.4 The g_m/I_D Sizing Methodology

The transconductance over drain current ratio is a resourceful tool for performing sizing.[2] The method exploits the fact that transconductances and drain currents vary like the gate width (as long as the widths are large enough to avoid border effects of course). Because the g_m/I_D ratio doesn't depend on the gate width, drain currents achieving any prescribed gain-bandwidth product can be derived from the expression below where the numerator is the transconductance given by Eq. 1.7 and the denominator the transistor's g_m/I_D ratio derived from a similar device whose gate width W^* and gate length L^* are known:

$$I_D = \frac{g_m}{\left(\frac{g_m}{I_D}\right)^*} \quad (1.17)$$

Knowing the drain currents, widths follow from the proportionality:

$$W = (W)^* \frac{I_D}{(I_D)^*} \quad (1.18)$$

Equations 1.17 and 1.18 form a set of parametric equations that determines drain currents and gate widths achieving the gain-bandwidth product fixed by g_m.[3] The key of the sizing methodology is the denominator of Eq. 1.17, for it plays the role of a parameter enabling to sweep the transistor through all modes of operation. It is actually the slope of the drain current characteristic plotted versus the gate voltage in semilog axes, for:

$$\left(\frac{g_m}{I_D}\right)^* = \frac{1}{I_D^*}\frac{dI_D^*}{dV_G} = \frac{d}{dV_G}\log(I_D^*) \quad (1.19)$$

---

[2] The gm/ID sizing methodology was introduced for the first time by the paper of (Silveira et al. 1996). Since then, the concept is referred to in many publication (Binkley et al. 2003; Binkley 2007) and (Girardi et al. 2006).

[3] When I_D and g_m are interchanged, sizing is aiming at slew-rate instead of the gain-bandwidth product.

In weak inversion, the slope of the drain current characteristic is large and practically constant. The currents derived from Eq. 1.17 are the smallest currents fulfilling the gain-bandwidth specifications. As we move towards strong inversion the slope decreases so that larger currents are needed to meet the gain-bandwidth specification.

The question is how to set up the denominator of Eq. 1.17? Two issues lie for the hand: experimental or analytical. The first makes use of Eq. 1.19 and derives the transconductance over drain current ratio from experimental $I_D(V_{GS})$ characteristics. The currents stored in look-up tables are the result of measurements carried out on real transistors whose width W^* and gate length L^* are known a priori. The drain currents may be derived also from advanced models such as BSIM or PSP[4] for these allow reconstructing drain currents that are very close to real drain currents. We call this the *semi-empirical g_m/I_D sizing method*. The other method, the *model-driven* method, makes use of analytical expressions for $(g_m/I_D)^*$. It requires having at one's disposal an accurate large signal model that lends itself to analytical expressions. The basic E.K.V. model introduced in Chapter 4 leads to analytic expressions of the transconductance over drain current ratio. Unfortunately, it is not a good candidate for it is too basic to take into consideration important second order effects like threshold voltage roll-off, D.I.B.L, gate length modulation etc that plague real MOS transistors. More elaborated version of the E.K.V. model (Enz and Vittoz 2006) do take care of these but evade chances to take advantage of analytic expressions owing to the large number of parameters and expressions they require. The semi-empirical method does not suffer of this drawback of course.

Yet, a simple model taking care of second order effects would be an asset. Further in this book, we show that when its parameters are not constant but vary with bias conditions and gate lengths, the basic E.K.V. model can be a good candidate nevertheless for model-driven sizing. Though the model itself ignores second order effects, the parameters reflect their impact. What makes this method attractive is the fact that analytic expressions offer sensible manners to control the mode of operation of the transistors, whereas the semi-empirical method proceeds blindly.

1.5 Conclusions

In this introductory chapter, we review the basics of sizing CMOS analog circuits. The transconductance over drain current ratio[5] offers a an interesting alternative for:

---

[4] BSIM is a widely used state-of-the-art model that is available in the public domain, see [BSIM]. It is based on threshold voltage formulations and this may explain some weaknesses in moderate inversion.

PSP for Penn State University and Philips (now NXP) is considered to be the more accurate industrial standard available nowadays (PSP 2006). It is based on the surface potential model (like the Charge Sheet Model).

[5] The method can be extended to other (trans)conductances. When the numerator and denominator of Eq. 1.17 are replaced respectively by g_d and g_d/I_D, the algorithm performs sizing in view of the output conductance.

1.5 Conclusions

- g_m/I_D is a technological attribute bridging the transconductance, a small signal quantity, to the drain current, a large signal quantity. As soon one is fixed, the other follows.
- The g_m/I_D ratio controls gain and power consumption, the larger g_m/I_D, the smaller the drain current and the larger the gain.
- The g_m/I_D sizing methodology applies however only as long as the widths are large enough to ignore lateral effects, a condition that holds true with most CMOS analog circuits.

Two approaches are possible: semi-empirical or model-driven. The first takes advantage of real measurements or data derived from advanced MOS models. The second makes use of models supposed to be accurate and simple enough to pave the way towards reliable analytical expressions of the transconductance over drain current ratios. Unfortunately no such model exists, except the basic E.K.V. model when its parameters are allowed to vary with bias conditions and gate lengths. We show further that the results are comparable to those obtained by means of the semi-empirical method.

Chapter 2
The Charge Sheet Model Revisited

2.1 Why the Charge Sheet Model?

We review in this chapter the main attributes of the 'Charge Sheet Model' (C.S.M.) introduced by J.R. Brews in 1978 (Brews 1978; Van de Wiele 1979). Although its name contains the word 'Model', the C.S.M is not a design tool. It is an invaluable means however for understanding some of the mechanisms governing current in MOS transistors for it scrutinizes phenomena otherwise difficult to apprehend. Unfortunately, the C.S.M. concerns only long channel MOS transistors implemented in a uniformly doped substrate (gradual channel approximation). Trying to predict drain currents of real transistors with the C.S.M. does not work.

Figure 2.1 depicts the structure of the NMOS transistor considered throughout this chapter. The two vertical lines without any other demarcation called respectively S and D symbolize the source and drain junctions. Two-dimensional effects are ignored, obliterating consequently items such as channel length modulation, Drain Induced Barrier Lowering (DIBL), etc. The source, drain and gate voltages are called respectively V_S, V_D and V_G, the surface potential ψ_S and the non-equilibrium voltage V.[1] The latter, called also the channel voltage, varies from V_S at the source to V_D at the drain. Single indices relate to voltages defined with respect to the substrate. Double indices relate to voltages defined with respect to references other than the substrate. For instance, V_{GS} is the voltage difference between the gate and the source.

2.2 The Generic Drain Current Equation

Current in MOS transistors results from mobile carries moving in the channel. It can be represented by the expression below where W is the width of the transistor and Q'_i the mobile charge density along the channel:

$$I_D = W \cdot (-Q'_i) \cdot velocity \qquad (2.1)$$

---

[1] V is the difference between the "quasi Fermi level" of electrons in the inversion layer and the "quasi Fermi level" of holes in the substrate.

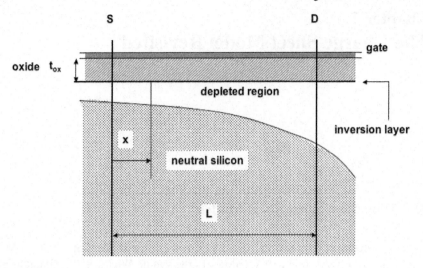

Fig. 2.1 The gradual channel MOS transistor

Two transport mechanisms are taking place currently: *drift* and *diffusion*. The *drift current* velocity is supposed to be proportional to the electrical field E:

$$drift\ current\ velocity = -\mu\, E \tag{2.2}$$

The *mobility coefficient* μ is assumed to be constant generally. This is correct as long as electrical fields do not exceed some limit. Modern transistors face very large fields for their gate lengths are ever shorter while supply voltages don't scale down necessarily at the same rate. As electrical fields are getting larger, the velocity of the carriers starts to slow down so that mobility declines. The effect can be taken into account by making μ a function of the electrical field.

The *diffusion current* is governed by the non-uniform concentration of carriers (like gas scattering in a closed vessel to homogenize the pressure). The diffusion current velocity is supposed to be proportional to the carrier's concentration:

$$diffusion\ current\ velocity = -D\frac{1}{n}\frac{\partial n}{\partial x} = -D\frac{1}{Q'_i}\frac{\partial Q'_i}{\partial x} \tag{2.3}$$

The *diffusion constant D* is related to the mobility μ by the Einstein relation:

$$D = \mu U_T \tag{2.4}$$

As the electrical field along the channel is replaced by the derivative of the surface potential ψ_S:

$$E = -\frac{d\psi_S}{dx} \tag{2.5}$$

Equation 2.1 can be rewritten as follows:

$$I_D = \mu W \left[-Q'_i \frac{d\psi_S}{dx} + U_T \frac{dQ'_i}{dx} \right] \quad (2.6)$$

or:

$$I_D \, dx = \mu W \left[-Q'_i \, d\psi_S + U_T \, dQ'_i \right] \quad (2.7)$$

While the left side of the above equation lends itself to integration (current is constant along the channel), the right part doesn't. One of the two integration variables should be expressed as a function of the other. Two strategies are possible. In the Charge Sheet Model, the charge density is expressed as a function of the surface potential. In the compact model, discussed in Chapter 4, the surface potential is expressed as a function of the charge density. The first representation follows a rigorous treatment while the second implies an approximation. The first does not lend itself to circuit design, the second does.

2.3 The Charge Sheet Model *Drain Current Equation*

In this chapter, we lay down the grounds of the Charge Sheet Model. We take the surface potential as integration variable rewriting the right part of Eq. 2.7 as shown below after introducing the gate oxide capacitance per unit-area C'_{ox}:

$$I_D \, dx = \mu C'_{ox} W \left[-\frac{Q'_i}{C'_{ox}} + U_T \frac{d}{d\psi_S} \left(\frac{Q'_i}{C'_{ox}} \right) \right] d\psi_S \quad (2.8)$$

To perform the integration, an expression of Q'_i / C'_{ox} versus the surface potential is required. The equation is derived currently from the total charge density Q_t / C'_{ox} expression obtained after combining the Gauss law, the Poisson equation and Boltzmann statistics (detailed computations can be found in textbooks):

$$-\frac{Q'_t}{C'_{ox}} = \gamma \cdot \left[U_T \exp\left(\frac{\psi_S - 2\phi_B - V}{U_T} \right) + \psi_S \right]^{1/2} \quad (2.9)$$

where:

V represents the non-equilibrium voltage along the channel

Φ_B is the bulk potential, depending on the ratio of the substrate doping concentration N over the intrinsic carrier density of silicon n_i

$$\phi_B = U_T \log\left(\frac{N}{n_i} \right) \quad (2.10)$$

γ is the Gamma commonly used in SPICE, which depends on N and the oxide thickness via the oxide capacitance C'_{ox}:

$$\gamma = \frac{1}{C'_{ox}} \sqrt{2q\varepsilon_S N} \qquad (2.11)$$

where q is the electron charge,
ε_S the silicon permittivity
N the substrate impurity concentration

The two terms under the square root of Eq. 2.9 relate respectively to the inversion charge density (left term) and the depleted charge density (right term). If we ignore the first term, in other words if the mobile charge density Q'_i vanishes, the total charge density Q'_t resumes to the fixed charge density Q'_b so that what remains of Eq. 2.9 boils down to:

$$-\frac{Q'_b}{C'_{ox}} = \gamma \sqrt{\psi_S} \qquad (2.12)$$

An expression of the mobile carrier's density lies now for the hand. We start from the Gauss law[2]:

$$V_G = -\frac{Q'_t}{C'_{ox}} + \psi_S \qquad (2.13)$$

Since Q'_t is the sum of mobile and fixed charge densities, we may write owing to Eq. 2.12:

$$V_G = -\frac{Q'_i}{C'_{ox}} + \gamma\sqrt{\psi_S} + \psi_S \qquad (2.14)$$

which leads to the expression of Q'_i/C'_{ox} versus the surface potential that we are looking for:

$$-\frac{Q'_i}{C'_{ox}} = V_G - \gamma\sqrt{\psi_S} - \psi_S \qquad (2.15)$$

We can evaluate now the derivative with respect to the surface potential of Q'_i/C'_{ox}:

$$d\left(-\frac{Q'_i}{C'_{oc}}\right) = -\left(1 + \frac{\gamma}{2\sqrt{\psi_S}}\right) d\psi_S \qquad (2.16)$$

and combine Eqs. 2.8, 2.15 and 2.16 to get the differential equation below ready for integration:

$$I_D dx = \mu C'_{ox} W \cdot \left[V_G - \gamma\sqrt{\psi_S} - \psi_S + U_T \left(1 + \frac{\gamma}{2\sqrt{\psi_S}}\right)\right] d\psi_S \qquad (2.17)$$

---

[2] The contact potentials between the gate material, the substrate material and the metal connections as well as the fixed charges trapped in the oxide produce a shift of the gate voltage that can be taken into account by adding to the gate voltage a constant voltage, called the Flat Band Voltage V_{FB}.

2.4 Common Source Characteristics

After integration, the expression of the drain current below is found where ψ_{SD} and ψ_{SS} represent respectively the surface potential at the drain and the source and β as usual $\mu C'_{ox} W/L$:

$$I_D = \beta \left[F(\psi_{SD}) - F(\psi_{SS}) \right] \tag{2.18}$$

with the function $F(\psi_S)$ given by:

$$F(\psi_S) = -\frac{1}{2}\psi_S^2 - \frac{2}{3}\gamma\,\psi_S^{1.5} + (V_G + U_T)\,\psi_S + \gamma\, U_T \psi_S^{0.5} \tag{2.19}$$

Equations 2.18 and 2.19 are interesting and frustrating results in the same time. The good news is that the drain current can be expressed as a polynomial of the square root of the surface potential. The bad news is that we must find a way to connect the source and drain surface potentials ψ_{SS} and ψ_{SD} to V_S and V_D. No analytical expression is available. The only way out is to extract the surface potential from the expression below resulting from the combination of Eqs. 2.13 and 2.9.

$$V_G = \gamma \cdot \left[U_T \exp\left(\frac{\psi_S - 2\phi_B - V}{U_T} \right) + \psi_S \right]^{1/2} + \psi_S \tag{2.20}$$

Since Eq. 2.20 is an implicit non-linear equation of ψ_S, the evaluation must be done numerically. The MATLAB function *surfpot* residing in the Matlab toolbox under 'extras.springer.com' takes care of this. For more details, please consult Annex 2.

2.4 Common Source Characteristics

The analytical expression of the drain current given by Eqs. 2.18 and 2.19 together with the *surfpot* function solving Eq. 2.20 pave the road towards experiments that help understanding the behavior of MOS transistors under low-power low-voltage conditions. Some examples are reviewed in the next sections considering an N-channel transistor implemented in a substrate having an impurity concentration equal to 10^{17} atoms cm^{-3} and an oxide thickness equal to 5 nm. The flat-band voltage V_{FB} is supposed to be equal to 0.6 V and the temperature equal to be 300 K. The reader can make use of the toolbox to run additional 'experiments', like those of Annex 3, which examine the impact of technology and temperature on transistor's performances.

2.4.1 The $I_D(V_D)$ Characteristics

The curves displayed in Fig. 2.2 show drain currents versus the drain voltage characteristics obtained by means of the program below making use of two additional MATLAB function, pMat and IDsh, reported in the toolbox. The unusual semilog

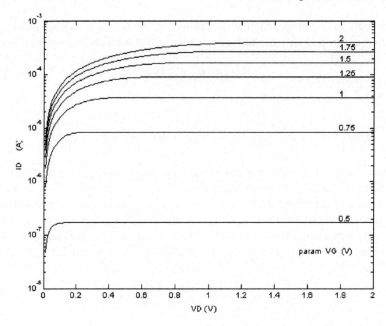

Fig. 2.2 Drain current versus drain voltage obtained by means of the *IDsh* file (MATLAB fig022.m)

vertical scale used for the display is chosen in order to plot drain currents from weak to strong inversion in a single diagram encompassing five orders of magnitude.

```
clear
clf
% data techno
T = 300;
N = 1e17;
tox = 5;
VFB = .6;

% compute pMat(technology vector)
p = pMat(T,N,tox);

% compute ID(VD)
VS = 0;
M = 201; VD = linspace(.01,2,M).';
UG = linspace (2,.5,7);
for k = 1: length(UG),
      ID(:,k) = IDsh(p,VS,VD,UG(1,k) + VFB);
end
% plot
semilogy(VD,ID,'k'); axis([0 2 1e-8 1e-3]);
```

2.4 Common Source Characteristics

A series of well-known facts are clearly visible:

1. The Charge Sheet Model represents drain currents in a smooth way all over the so-called linear (resistive) and saturated modes of operation. The model is 'continuous'. In other words it does not require several equations to describe distinct modes of operation.
2. The passage from strong to weak inversion and vice – versa is gradual and continuous too.
3. The distances between adjacent $I_D(V_D)$ characteristics gets larger as one goes from strong to weak inversion. Since all gate voltage increments are identical, the transconductance over drain current ratio is larger in weak than in strong inversion (remind weak and moderate inversion conditions achieve better gains).
4. The pinch-off voltage is very small in weak inversion and remains quasi-constant throughout the weak inversion regime. It is of the order of 100 mV, similar to the saturation voltage of bipolar transistors.
5. Drain currents in saturation are quasi-constant for the C.S.M. ignores effects like channel length modulation and DIBL. The transistor behaves like a perfect current source.

2.4.2 The $I_D(V_G)$ Characteristic of the Saturated Transistor

Figure 2.3 shows the drain current versus the gate voltage of the same transistor as above when saturated (the drain voltage V_D has no influence on the drain current). The almost linear section left attests clearly that below 0.5 V (weak inversion) the

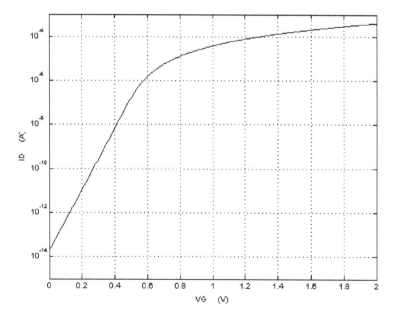

Fig. 2.3 Drain current of the saturated common source transistor (MATLAB fig023.m)

drain current increases quasi-exponentially. In this region, the *'subthreshold'* slope S determines the so-called slope factor n:

$$n = \frac{S}{U_T \log(10)} \quad (2.21)$$

Beyond the exponential, the drain current levels off gradually while the transistor is entering moderate and strong inversion. The trend in the strong inversion region is quadratic.

2.4.3 Drift and Diffusion Contributions to the Drain Current

The C.S.M. offers the possibility to compare the contributions of drift and diffusion currents to the drain current. All what is needed therefore is to break the polynomial representation of I_D of Eq. 2.19 into two parts. For the diffusion current, the two last terms of Eq. 2.17 are considered and for the drift current what remains. The polynomials are respectively:

$$P_{\mathit{diffusion}} = \begin{bmatrix} 0 & 0 & U_T & \gamma U_T & 0 \end{bmatrix} \quad (2.22)$$

and

$$P_{\mathit{drift}} = \begin{bmatrix} -\frac{1}{2} - \frac{2}{3}\gamma & V_G & 0 & 0 \end{bmatrix} \quad (2.23)$$

The drift and diffusion currents displayed in Fig. 2.4 show clearly the dominance of one current over the other depending on which mode of operation is taking the lead. Diffusion dominates in weak inversion while drift takes over in strong inversion. In the middle, around 0.6 V, drift and diffusion currents have almost the same magnitudes. In strong inversion, the total current coincides practically with the quadratic approximation of I_D. The same holds true for the exponential current in weak inversion. Many analog circuits, especially low-power low-voltage circuits, operate nowadays in the so-called moderate inversion region.

2.5 Weak Inversion Approximation of the Charge Sheet Model

The fact that diffusion current overwhelms drift current in weak inversion leads to a number of useful approximate expressions. Because the first of the two right terms of Eq. 2.6 can be ignored, one has:

$$I_D \, dx \approx -\mu \, C'_{ox} W \, U_T \, d\left(-\frac{Q'_i}{C'_i}\right) \quad (2.24)$$

2.5 Weak Inversion Approximation of the Charge Sheet Model

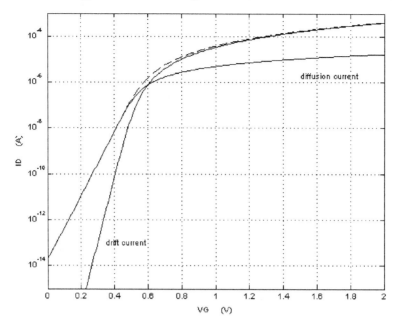

Fig. 2.4 Same as Fig. 2.3 with explicit representations of the drift and diffusion currents (MATLAB fig024.m)

Consequently, the drain current in weak inversion is given by:

$$I_D \approx -\mu \, C'_{ox} \frac{W}{L} \, U_T \left[\left(-\frac{Q'_{iD}}{C'_{ox}} \right) - \left(-\frac{Q'_{iS}}{C'_{ox}} \right) \right] \quad (2.25)$$

Let us find now an approximate expression of Q'_i/C'_{ox} in weak inversion. The inversion layer charge density is extracted from the equality:

$$-\frac{Q'_i}{C'_{ox}} = -\frac{Q'_t}{C'_{ox}} + \frac{Q'_b}{C'_{ox}} \quad (2.26)$$

where Q'_t/C'_{ox} and Q'_b/C'_{ox} are replaced by Eqs. 2.9 and 2.12. This leads to:

$$-\frac{Q'_i}{C'_{ox}} = \gamma \cdot \left[U_T \exp\left(\frac{\psi_S - 2\phi_B - V}{U_T} \right) + \psi_S \right]^{1/2} - \gamma \sqrt{\psi_S} \quad (2.27)$$

Since the contribution of the first of the two terms under the square root (drift current) is small compared to that of the second (diffusion current), the equation above can be approximated as follows:

$$-\frac{Q'_i}{C'_{ox}} = \gamma \sqrt{small + \psi_S} - \gamma \sqrt{\psi_S} \approx \gamma \frac{small}{2\sqrt{\psi_S}} \quad (2.28)$$

This leads to:

$$-\frac{Q'_i}{C'_{ox}} \approx \gamma \frac{U_T}{2\sqrt{\psi_S}} \exp\left(\frac{\psi_S - 2\phi_B}{U_T}\right) \cdot \exp\left(-\frac{V}{U_T}\right) \quad (2.29)$$

The surface potential ψ_S depends practically only on the gate voltage. In weak inversion, Eq. 2.14 boils down indeed to a second order equation relating the gate voltage V_G to ψ_S for Q'_i/C'_{ox} is small in comparison to the contribution of the depletion layer represented by the two last terms. An expression of the weak inversion surface potential ψ_{Swi} can be extracted then from the latter:

$$\psi_{Swi} = \left[-\frac{\gamma}{2} + \sqrt{\left(\frac{\gamma}{2}\right)^2 + V_G}\right]^2 \quad (2.30)$$

When ψ_{Swi} is put in Eq. 2.29 and the latter combined with Eq. 2.25, the next expression of the drain current in weak inversion is obtained:

$$I_D \approx \underbrace{\frac{1}{2}\beta\gamma U_T^2 \frac{1}{\sqrt{\psi_{Swi}}} \exp\left(\frac{\psi_{Swi} - 2\phi_B}{U_T}\right)}_{A} \cdot \left[\exp\left(-\frac{V_S}{U_T}\right) - \exp\left(-\frac{V_D}{U_T}\right)\right]$$

$$(2.31)$$

This is an interesting result for it shows that the drain current in weak inversion is controlled exponentially by the source and drain voltages owing to the fact that the factor A depends only on the gate voltage. Another interesting observation concerns the drain voltage when the transistor enters saturation. Rewriting Eq. 2.31 in terms of the drain-to-source voltage difference V_{DS} turns the above expression into:

$$I_D \approx A \cdot \exp\left(-\frac{V_S}{U_T}\right) \cdot \left[1 - \exp\left(-\frac{V_{DS}}{U_T}\right)\right] \quad (2.32)$$

The drain current saturates as soon as the drain-to-source voltage attains 100 mV (nearly four times U_T). The impact of the gate voltage is more difficult to apprehend for it is hidden in the A factor, which depends on the geometry via β, and γ, ϕ_B and the surface potential. The point is discussed more in detail further.

2.6 The g_m/I_D Ratio in the Common Source Configuration

An analytic expression of the transconductance over drain current ratio cannot be derived from the Charge Sheet Model. The ratio must be evaluated numerically by taking the derivative with respect to the gate voltage of the log of the drain current:

$$\frac{g_m}{I_D} = \frac{1}{I_D}\frac{\partial I_D}{\partial V_G} = \frac{\partial \log(I_D)}{\partial V_G} \quad (2.33)$$

2.6 The g_m/I_D Ratio in the Common Source Configuration

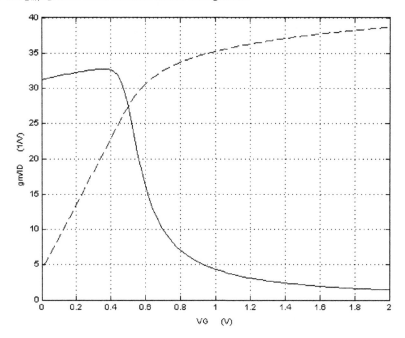

Fig. 2.5 The g_m/I_D ratio derived from the $I_D(V_G)$ plot of Fig. 2.4, which is reproduced in the background (MATLAB fig025.m)

The g_m/I_D ratio is shown in Fig. 2.5. It is larger in weak than in strong inversion but displays a slightly decaying trend under very small currents. The phenomenon can be explained as follows. Consider the derivative versus the gate voltage of the expression hereafter extracted from Eq. 2.31:

$$\left(\frac{g_m}{I_D}\right)_{W.I.} \approx \frac{\partial}{\partial V_G} \log\left[\frac{1}{\sqrt{\psi_{Swi}}} \exp\left(\frac{\psi_{Swi} - 2\phi_B}{U_T}\right)\right] \quad (2.34)$$

The derivative is done into two steps, first with respect to V_G, second to ψ_{Swi}:

$$\left(\frac{g_m}{I_D}\right)_{W.I.} \approx \left(-\frac{1}{2\psi_{Swi}} + \frac{1}{U_T}\right) \frac{\partial \psi_{Swi}}{\partial V_G} \quad (2.35)$$

When the derivative of the weak inversion surface potential with respect to the gate voltage is extracted from Eq. 2.30, one has:

$$\left(\frac{g_m}{I_D}\right)_{W.I.} = \frac{1}{U_T} \cdot \frac{1 - \frac{U_T}{2\psi_{Swi}}}{1 + \frac{\gamma}{2\sqrt{\psi_{Swi}}}} \quad (2.36)$$

Notice the similarity with the approximate g_m/I_D ratio of Eq. 1.13 in the first chapter, stating that:

$$\left(\frac{g_m}{I_D}\right)_{W.I.} = \frac{1}{n\,U_T} \qquad (2.37)$$

The comparison of Eq. 2.36 with 2.37 brings about an interesting analytical expression of the weak inversion subthreshold slope factor, called n_{wi}:

$$n_{wi} = \frac{1 + \frac{\gamma}{2\sqrt{\psi_{Swi}}}}{1 - \frac{U_T}{2\psi_{Swi}}} \qquad (2.38)$$

The signification of n_{wi} gets clear when several g_m/I_D plots are merged as their source voltages changes. Figure 2.6 shows clearly that when the transistor enters weak inversion, all g_m/I_D ratios come together forming a single consolidated envelope, which coincides with Eq. 2.36.

Near the origin, the envelope bends down more rapidly for the width of the depleted region under the gate is getting smaller as the source voltage decreases. The ratio of the capacitive divider formed by the gate oxide and the depleted region increases, modifying consequently the slope factor n_{wi}.

One may substitute a more compact and more familiar expression to Eq. 2.31:

$$I_D \approx \underbrace{I_o \exp\left(\frac{V_G}{n_{wi}\,U_T}\right)}_{A} \cdot \left[\exp\left(-\frac{V_S}{U_T}\right) - \exp\left(-\frac{V_D}{U_T}\right)\right] \qquad (2.39)$$

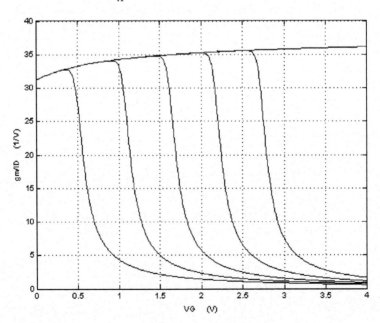

Fig. 2.6 Plot of g_m/I_D ratios when the source voltage changes from 0 V (*left*) to 2 V (*right*) in steps 0.5 V wide. The distance separating g_m/I_D's is a little more than 0.5 V owing to the body-effect discussed more in detail in the next chapter (MATLAB fig026.m)

2.7 Common Gate Characteristics of the Saturated Transistor

Let us consider now the common-gate configuration. The gate voltage is fixed while the source V_S is the input now. Figure 2.7 shows the $I_D(V_S)$ curve obtained after running the same file as above when the gate voltage is equal to 2 V and the source voltage V_S varies from 0 to 2 V. The drain voltage is supposed to be large enough in order to keep the transistor saturated under any circumstance. As V_S increases, the gate-to-source voltage V_{GS} decreases abating the drain current. First, the drain current slows down gradually for the transistor is still in strong inversion. Once the transistor is in weak inversion, the current decreases exponentially. In this region, the slope of the drain current follows the $\exp(V_S/U_T)$ law predicted by Eq. 2.31. Put differently, the slope factor is equal to one in strong contrast with the slope factor of the common source slope factor.

The same plot shows also the g_{ms}/I_D ratio inferred from the drain current characteristic. Like in the common-source configuration, the ratio is given by the slope of the semilog-scaled drain current. The 'transconductance' is now g_{ms} instead of g_m. The sequence is being reversed with respect to the common-source configuration for the g_{ms}/I_D is flipped horizontally with respect to g_m/I_D.

In weak inversion, the g_{ms}/I_D ratio is equal to $1/U_T$ in accordance with the unity slope factor mentioned above. This once again underlines the similarity MOS

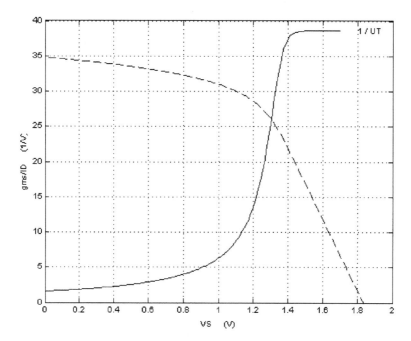

Fig. 2.7 The g_{ms}/I_D ratio (*plain lines*) versus the source voltage is obtained by taking the derivative of the log scaled drain current plotted in *dashed lines* (V_G is equal to 2 V) (MATLAB fig027.m)

transistors share with bipolar transistors when operating in weak inversion. Two reasons explain this. First, the drain current is dominated by diffusion current as in the neutral base of the bipolar transistor. Second, the front and back gates cooperate whereas the back-gate remains idle in the common source configuration. The common-gate configuration ignores consequently the partitioning inherent to the common-source configuration.

We will show in later chapters that real transistors do not conform to the unity slope factor in weak inversion in the common-gate configuration. The slope factor is generally slightly larger than one. This is due to the drain-to-source voltage variation going along with the gate-to-source voltage modifications. In the C.S.M. the drain voltage has no effect on the current as long as the transistor is saturated, but with real transistors, changes of the drain voltage modify the space charge near the drain and below the inversion layer. The drain influences thus the current even though the transistor is saturated. The g_{ms}/I_D ratio of real transistors in weak inversion is smaller thus than the predicted $1/U_T$.

2.8 A Few Concluding Remarks Concerning the C.S.M.

The Charge Sheet Model is a physical model that predicts drain currents whatsoever mode of operation, weak or strong inversion, saturation or not. It is relevant and particularly instrumental for understanding the basic mechanisms controlling low-power operation. In addition, the model bridges drain currents to physical quantities such as the substrate impurity concentration, oxide thickness and temperature. It offers therefore the possibility to scrutinize sensitivity aspects. The validity of the C.S.M. is restricted however to ideal transistors implemented in a uniformly doped substrate with gate lengths sufficiently large to obliterate short channel effects.

An interesting observation can be made as far as the *threshold voltage*. So far, the concept has not been mentioned except occasionally, for instance when the quadratic model was considered. The Charge Sheet Model ignores actually the concept. The reason is that the threshold voltage is not a physical quantity but a parameter embodied on measurements. Its interpretation varies according to the evaluation techniques. This does not mean that the threshold voltage is a useless concept. On the contrary, it is a landmark, like the voltage drop across forward biased junctions. It is an essential parameter exploited in every circuit oriented model. In the next chapter, we are going to clarify the concept.

Chapter 3
Graphical Interpretation of the Charge Sheet Model

3.1 A Graphical Representation of I_D

An interesting representation of the drain current can be obtained when the expression below is used for the drain current (Tsividis 1999):

$$I_D = \mu C'_{ox} \frac{W}{L} \cdot \int_{V_S}^{V_D} \left(-\frac{Q'_i}{C'_{ox}}\right) dV \tag{3.1}$$

The equation is derived from, the proportionality of the minority carrier density to the exponential function acknowledged by Boltzmann statistics:

$$Q'_i \propto \exp\left(\frac{\psi_S - 2\phi_B - V}{U_T}\right) \tag{3.2}$$

After differentiating the two sides of the above expression, an equation connecting the differentials of Q'_i, ψ_S and V is obtained:

$$U_T \frac{dQ'_i}{Q'_i} = d\psi_S - dV \tag{3.3}$$

This enables us to replace the drift and diffusion current contributions considered in the previous chapter by a single term, $Q'_i \, d\psi_S$, turning Eq. 2.7 into the expression below, which leads to Eq. 3.1 after integration:

$$I_D \, dx = \mu C'_{ox} W \left(-\frac{Q'_i}{C'_{ox}}\right) dV \tag{3.4}$$

Although Eq. 3.1 is more compact than Eq. 2.8, we haven't booked any progress for the integration has to be carried out now with respect to the channel voltage V while the expression between brackets is a function of the surface potential ψ_S as reminded by Eq. 2.15 reproduced hereunder for convenience:

Fig. 3.1 Representation versus the non-equilibrium voltage V of the surface potential ψ_S and the threshold voltage with respect to the substrate V_T. The p vector is given by *pMat*(300,1e17,5) and the gate voltage equal to 2 V (MATLAB fig031.m)

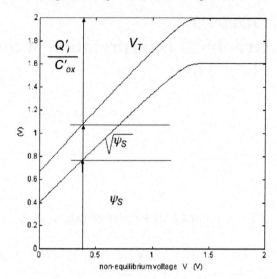

$$-\frac{Q'_i}{C'_{ox}} = V_G - \gamma\sqrt{\psi_S} - \psi_S \qquad (3.5)$$

Equations 3.4 and 3.5 pave the road however towards a graphical interpretation of the drain current. The idea is illustrated by the two curves shown in Fig. 3.1: the lower one representing the surface potential ψ_S versus the non-equilibrium voltage V, the upper one, called V_T, the sum hereunder:

$$V_T = \gamma\sqrt{\psi_S} + \psi_S \qquad (3.6)$$

According to Eqs. 3.5 and 3.6, V_T is the voltage to apply to the gate in order to zero the mobile charge density Q'_i. When V_G is larger than V_T, the semiconductor surface is inverted and when V_G is smaller than V_T there is no inversion layer. Hence, V_T can be assimilated to a kind of **threshold voltage**, which should not be confused with the threshold voltage V_{th} currently associated to the quadratic representation of the drain current. The first is defined with respect to the substrate, the second with respect to the source.

Because the difference between V_G and the threshold voltage V_T is in a representation of $-Q'_i/C'_{ox}$, we may rewrite Eq. 3.1 as follows:

$$I_D = \mu C'_{ox}\frac{W}{L} \cdot \int_{V_S}^{V_D} (V_G - V_T)\,dV \qquad (3.7)$$

3.1 A Graphical Representation of I_D

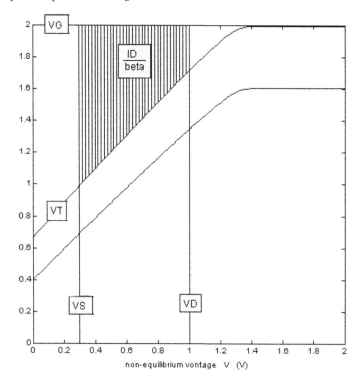

Fig. 3.2 Graphical illustration of the drain current of a MOS transistor whose V_S and V_D are respectively equal to 0.3 and 1.0 V. The gate voltage, oxide thickness, substrate doping and temperature are the same as in Fig. 3.1 (MATLAB fig032.m)

This leads to the graphical interpretation of the drain current[1] represented by the hatched area of Fig. 3.2. The surface delineated by V_G and V_T and the vertical lines V_S and V is indeed a representation of the drain current divided by β according to Eq. 3.7. This representation can be used in order to visualize how the terminal voltages control the drain current. Consider for instance a grounded source transistor whose drain voltage V_D increases gradually, starting from zero. When V_D is small, the area representing the drain current divided by beta resolves to a narrow stripe very close to the vertical axis like in the first of the four views shown in Fig. 3.3. As the drain voltage increases, the area widens but the growth rate declines as we approach the point where V_T gets close to V_G. Beyond this point, the drain current does not increase anymore for the triangularly shaped area representing the drain current remains practically constant. We reached the ***pinch-off voltage***. The transistor is now saturated.

---

[1] The graphical interpretation of the drain current is designated generally by the name their authors Memelink–Jespers. It was reported first in (Jespers et al. 1977) and taken over in a number of publications among which Cand et al. (1986), Wallinga and Bult (1989) and Enz and Vittoz (2006).

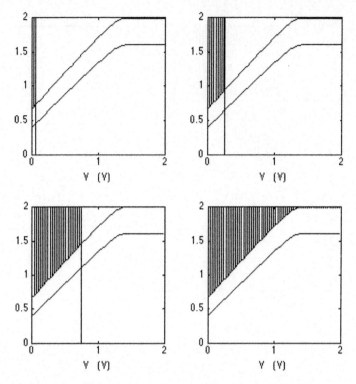

Fig. 3.3 Graphical illustration of $I_D(V_D)/beta$ considering a grounded source transistor with a constant V_G of 2 V and a drain voltages stepping from 0.050 V, to 0.250 V, 0.750 V and 2,00 V (MATLAB fig033.m)

3.2 More on the V_T Curve

Before we illustrate by means of a few examples the use that can be made of the graphical construction, we look more closely to the surface potential curve shown in Fig. 3.1 in order to explain its shape.

Figure 3.4 shows two representations of ψ_S. The left one traces the surface potential ψ_S versus the gate voltage V_G considering a series of constant non-equilibrium voltages V increments from 0 to 2 V in steps of 0.5 V. The right figure shows similar data, plotted versus the non-equilibrium voltage V.

Left, near the origin, all the surface potential curves merge. Soon breakpoints appear beyond which ψ_S remains quasi constant. Breakpoints shift to larger gate voltages as V increases. Left to every breakpoint the surface is not inverted. The width of the depleted region is widening in order to balance the gate charge as V_G increases. The non-linear capacitive divider formed by the gate oxide and the depleted region determines the surface potential. Right to the breakpoints, mobile charges start accumulating along the semiconductor surface. The width of the depleted region doesn't change anymore for the charge represented by the mobile

3.3 Two Approximate Representations of V_T

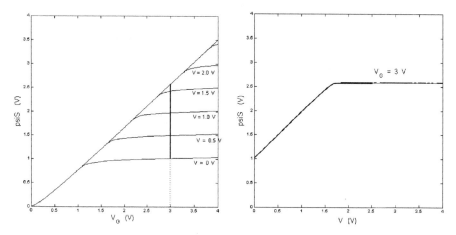

Fig. 3.4 ψ_S versus V_G (*left*) and V (*right*) (MATLAB fig034.m)

carriers increases almost at the same rate as the gate charge. The surface potential does not change either. Of course, larger gate voltages are needed to invert the surface as V increases. In the right figure, the roles of the gate voltage and the non-equilibrium voltage are interchanged. The surface potential is plotted versus the non-equilibrium voltage V while the gate voltage V_G is kept constant. Consider for instance a V_G equal to 3 V, corresponding to the vertical line of the left plot. As we move up along this line, the charge in the inversion layer decreases while the width of the depleted region increases in order to balance the more or less constant gate charge. When the point is reached where the vertical line meets the depleted characteristic, the inversion charge vanishes. The surface potential remains quasi-constant notwithstanding the fact that V keeps on growing, the excess charge being taken over by the widening depleted region.

3.3 Two Approximate Representations of V_T

The strong resemblance to a broken line of the surface potential in the right part of Fig. 3.4 legitimates the introduction of approximations. These lead to two well-known expressions of I_D that are reviewed briefly hereunder.

3.3.1 The 'Linear' Surface Potential Approximation

Since the slope of the surface potential below pinch-off remains almost constant, ψ_S may be approximated by means of a linear expression:

$$\psi_S = \psi_{So} + V \tag{3.8}$$

ψ_{So} being the surface potential at the origin, generally equal to $2\Phi_B$ plus k times U_T, k being comprised between 4 and 8:

$$\psi_{So} = 2\Phi_B + kU_T \qquad (3.9)$$

The threshold voltage V_T with respect to the substrate defined by Eq. 3.6 is then given by:

$$V_T = \gamma\sqrt{\psi_{So} + V} + \psi_{So} + V \qquad (3.10)$$

which can be rewritten as follows:

$$V_T = V_{To} + \gamma\left(\sqrt{\psi_{So} + V} - \sqrt{\psi_{So}}\right) + V \qquad (3.11)$$

Plugging this expression in Eq. 3.7 leads to the well-known expression of the drain current below:

$$I_D = \beta\left[(V_G - \psi_{So})V - \frac{2}{3}\gamma(V_G + \psi_{So})^{1.5} - \frac{1}{2}V^2\right]_{V_S}^{V_D} \qquad (3.12)$$

Figure 3.5 shows the drain current predicted by Eq. 3.12 considering various values of k and a gate voltage of 3 V (no flat band voltage correction). Below saturation, the

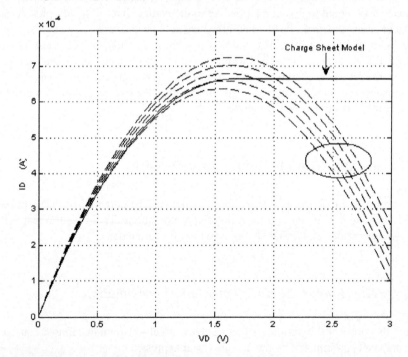

Fig. 3.5 Comparison of the drain current evaluated by means of Eq. 3.12 to the current predicted by the Charge Sheet Model. The parameter k illustrated by the curves in the ellipse varies from 4 to 8 by steps equal to 1 (MATLAB fig035.m)

3.3 Two Approximate Representations of V_T

current reproduces nicely the current predicted by the C.S.M. transistor of Chapter 2 when k is equal to 7. Beyond the maximum, the drain current should not drop of course but remain constant. The difference comes from the linear approximation of V_T, which should break away instead of growing above the pinch-off voltage. While more or less correct in strong inversion, Eq. 3.12 does not represent the reality however in moderate and weak inversion for the abrupt change of V_T that occurs at the pinch-off voltage departs strongly from the smooth passage conveyed by the Charge Sheet Model.

3.3.2 The 'Linear' Threshold Voltage V_T Approximation

Below pinch-off, we can approximate the threshold voltage by a linear expression. The approximate expression of V_T is given then by the equation below where the slope factor n is supposed to be constant and slightly larger than one:

$$V_T = V_{To} + nV \tag{3.13}$$

Of course, we should expect a larger error for V_T sums up not only of the quasi-linear surface potential but also γ times the square root of the surface potential. This turns Eq. 3.7 into the well-known 'quadratic' drain current equation:

$$I_D = \beta \left[(V_G - V_{To}) V - \frac{n}{2} V^2 \right]_{V_S}^{V_D} \tag{3.14}$$

The point where V_T crosses V_G, the pinch-off point, is given now by the well-known expression:

$$V_P = \frac{V_G - V_{To}}{n} \tag{3.15}$$

The graphical representation of the drain current illustrated by Fig. 3.3 is now very simple. The plot representing V_T resumes to a broken line consisting of a straight line with a slope n crossing the vertical axis at the threshold voltage V_{To}, which turns into a horizontal line at the pinch-off voltage when V_T equals V_G. When V_D is smaller than the pinch-off voltage V_P, the area boils down to the difference between a rectangle and a triangle. The area of the rectangle is equal to $(V_G - V_{To})$ times V_D and the area of the triangle given by $nV_D^2/2$. When V_D is larger than V_P, the current is given by $(V_G - V_{To}).V_P/2$ or $(V_G - V_{To})^2/2n$. These reproduce the well-known quadratic drain currents expressions after multiplication by β.

3.4 A Few Examples Illustrating the Use of the Graphical Construction

In the sections hereafter, we review a series of examples illustrating the use that can be made of the graphical construction. We consider for V_T the last linear approximation.

3.4.1 The MOS Diode

The first example is given by the diode-connected common source MOS transistor shown in Fig. 3.6. Since V_G is equal to V_D, the vertical line representing V_D and the horizontal line representing V_G cross each other on the dashed line dividing the square in two equal parts. The graphical counterpart of the drain current boils down then to the triangle entangled between the vertical axis (V_S is equal to zero), the horizontal line representing the gate voltage V_G and the threshold voltage V_T. As V_G is lifted up, the area – or the current – grows quadratically. It is clear that diode-connected MOS transistors are always saturated, whatsoever the drain current for the pinch-off voltage (the point where V_G crosses V_T) lies always below V_D.

3.4.2 The MOS Source Follower

Figure 3.7 represents a source follower fed by a constant current source I_D. The transistor is supposed to be saturated. Since the current is fixed, the area of the triangle representing the drain current remains constant. The horizontal side of the triangle is fixed by V_G as usual while the vertical corresponding to the source voltage V_S is fixed by the current, thus the area. When V_G changes, the triangle glides along V_T causing a concomitant shift of the source voltage. The ratio ΔV_S over

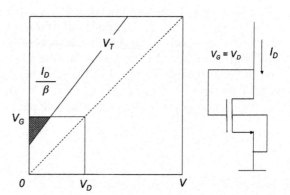

Fig. 3.6 Graphical illustration of the current-voltage relation of a MOS diode

3.4 A Few Examples Illustrating the Use of the Graphical Construction 33

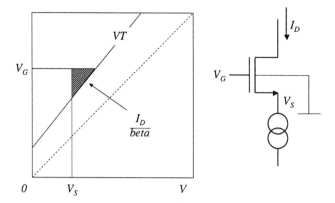

Fig. 3.7 Graphical illustration of the input-output relation the MOS source-follower

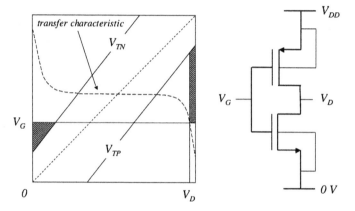

Fig. 3.8 Graphical illustration of the input-output characteristic of a CMOS inverter

ΔV_G, which represents the gain of the source follower, is equal to the reciprocal of the slope factor n confirming the well-known fact that MOS source followers have gains always smaller than one by 20–40%.

3.4.3 The CMOS Inverter

Logical inverters combine N and P-type MOS transistors. Two V_T's must be considered instead of one thus. The substrate is the reference of the N-MOS transistor and the N well connected to the power supply V_{DD} the reference of the P-MOS transistor. The lower-left corner of the plot of Fig. 3.8 is thus the origin of axes for the N-MOS transistor while the upper-right corner is the origin for the P-MOS transistor. The V_T's of the two transistors are shown respectively. Since the gates are shorted, the horizontal lines representing the gate voltages of both transistors are merged. Similarly, the drain voltages of the N- and P-type transistors are represented by means of a single vertical line. The areas representing the currents of the N- and P-type transistors are now sketched. They are proportional naturally for the same current is flowing in the two transistors. If the W/L's are sized in order to compensate the unfavorable mobility ratios of holes over electrons, the areas must be equal. The graphical construction boils down then to a simple geometrical problem: find the drain voltage that makes the hatched areas equal. When V_G is low, the area of N-type transistor confines to a small triangle. The only way to equalize areas is to shift the vertical line very close to V_{DD}. The N-channel transistor is saturated while the P-channel isn't. When V_G is large, the opposite holds true.

Since the gate and drain terminals represent respectively the input and the output of the logic inverter, the intersection of the V_G and V_D lines reproduces the inverter I/O transfer characteristic after flipping horizontal and vertical axes. Since the slope along the transfer characteristic represents the small signal gain of the inverter, the gain grows as we move towards the centre until V_G gets equal to half the power supply (assuming both transistors have identical threshold voltages). The currents in the N- and P-channel transistors are then represented by means of two identical triangles, meaning that both transistors are saturated. Any V_D between the two pinch-off voltages is then a plausible output voltage. The small signal gain is thus infinite! Of course, this is not correct. The errors comes from the fact that the construction assumes the output conductance of saturated transistors is equal to zero. Like in the Charge Sheet Model, the construction does not take into account second order effects, like the Early effect.

3.4.4 Small Signal Transconductances

Besides large signal quantities, the graphical construction helps also to 'visualize' small signal parameters like g_m or g_{ms}. What is needed therefore is to consider small changes of the drain current illustrated by small departures of the lines representing V_G or V_S. In Fig. 3.9, we consider the drain current changes resulting from gate voltage variations:

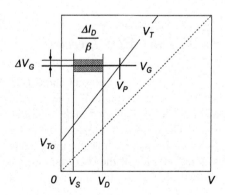

Fig. 3.9 Graphical evaluation of g_m

3.4 A Few Examples Illustrating the Use of the Graphical Construction

$$\Delta V_G(V_D - V_S) = \frac{\Delta I_D}{\beta} \tag{3.16}$$

This implies that:

$$\frac{g_m}{\beta} = (V_D - V_S) \tag{3.17}$$

and, when the transistor is saturated:

$$\frac{g_m}{\beta} = (V_P - V_S) \tag{3.18}$$

The length of the segment comprised between V_S and V_D is the graphical counterpart thus of the transconductance g_m divided by β.

Similarly, when the vertical line representing V_S moves slightly around its steady state position, one has:

$$\frac{g_{ms}}{\beta} = (V_G - V_T) \tag{3.19}$$

We see that when the transistor is saturated, the ratio of the source transconductance over gate transconductance is equal to n, confirming a statement made earlier in Chapter 2:

$$\frac{g_{ms}}{g_m} = \frac{V_D - V_S}{V_G - V_T} = n \tag{3.20}$$

3.4.5 CMOS Transmission Gates

One can make use of the graphical construction in order to explain why some circuits are preferred to others. For instance: why are digital CMOS transfer gates implemented by means of parallel complementary transistors rather than by single transistors? Figure 3.10 compares the conductance of a single transistor to that of a complementary switch. The upper part of the figure relates to the single transistor, which is supposed to be connected between a voltage source and a load capacitor. The current is equal to zero for we assume that steady state conditions are attained. The area representing the drain current resumes to a segment whose length represents the conductance of the switch divided by β. It is obvious that so-called 'dead zones' occur for some input voltages. Two MOS transistors of opposite types in parallel like in the lower part of the same figure do not suffer from the same impairment. The conductance of the transmission gate is the sum of two segments. As one is vanishing, the other is taking over. There is no 'dead zone'.

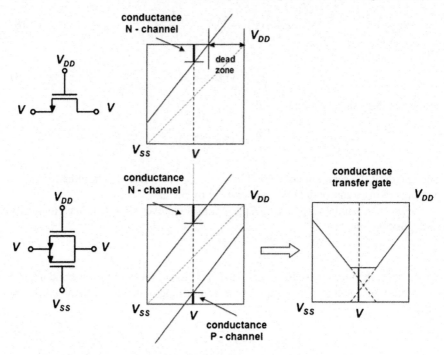

Fig. 3.10 The digital 'transmission gate'

3.4.6 How to Implement Quasi-linear Resistors with MOS Transistors

Continuous filters make use currently of integrated resistors and capacitors. Quasi-linear capacitors are normally available in MOS technology but resistors require dedicated circuits. The circuit of Fig. 3.11 shows an implementation of a quasi-linear resistor. The gate of the MOS transistor is connected to a constant bias voltage V_G while the source and drain undergo equal and opposite voltage excursions with respect to a constant reference voltage V_o called DV. If V_T is assimilated to a linear function of V, it is clear that the trapezoidal hatched area is equal to the area of the rectangle with thick lines. The drain current depends linearly on DV thus.

V_T however is a slightly quadratic function of V and this impairs the resistor's linearity. A better 'resistor' proposed by (Banu and Tsividis 1984) is shown in Fig. 3.12.

The circuit consists of two MOS 'resistors' with common sources and drains but distinct gate voltages. When the current delivered by one 'resistor' is subtracted from the current delivered by the other, the non-linearity associated with V_T doesn't impair performances any more. The area of the rectangle representing the difference of the two currents is independent of V_T (Wallinga and Bult 1989). In practice, non-linear distortion decreases substantially but doesn't disappear. Another cause

3.4 A Few Examples Illustrating the Use of the Graphical Construction

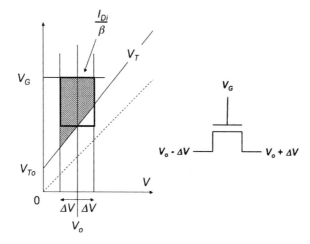

Fig. 3.11 A MOS transistor implementing a 'resistor'

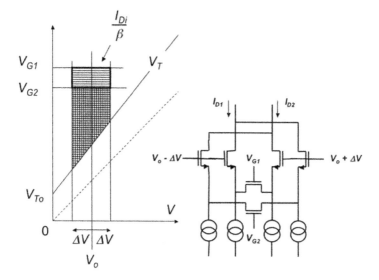

Fig. 3.12 Implementation of a quasi-linear resistor

of imperfection still remains: the vertical electrical fields in the two transistors are not identical. Mobility mismatch is now the prime source of non-linear distortion.

3.4.7 Source-Bootstrapping

Dynamic circuits can produce voltages that are larger than the supply voltage. The way such circuits operate can be illustrated intuitively by the graphical construction. The source bootstrap circuit shown in Fig. 3.13 offers a typical example.

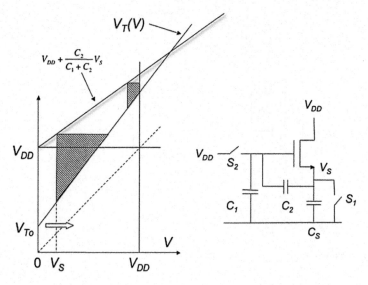

Fig. 3.13 Source bootstrapping

The circuit consists of a MOS transistor whose source is connected to ground through a capacitor C_S in parallel with a switch S_1. The gate is connected to the supply voltage through another switch S_2. When both switches open, current starts charging C_S. As the source voltage increases, it pushes the gate voltage up through the capacitive divider formed by C_2 and C_1. The gate voltage increase is actually an attenuated replica of the source voltage. As the source voltage increases, the triangle representing the drain current divided by β moves from left to right squeezed between the gate voltage and the threshold voltage V_T lines. The area of the triangle decreases gradually until the point is reached where the transistor begins to de-saturate. Finally, the source attains V_{DD} and the current is zeroed. The gate voltage is then equal to $V_{DD}{}^*(C_1 + 2C_2)/(C_1 + C_2)$. Because the slope of the line representing the gate voltage is smaller than that of V_T, there is point beyond which the gate voltage cannot go.

This illustrates clearly the performance limitations caused by the substrate effect or the slope factor n. The same construction can be used in order to illustrate the functioning of Bucket-Brigade Devices.

3.5 A Closer Look to the Pinch-Off Region

So far, all the examples we considered concern strong inversion. They don't show what is happening near the pinch-off voltage. When the transistor enters moderate and weak inversion, the approximation representing V_T by means of a broken line is too basic. A more correct image of the surface potential near pinch-off is shown

3.6 Conclusion

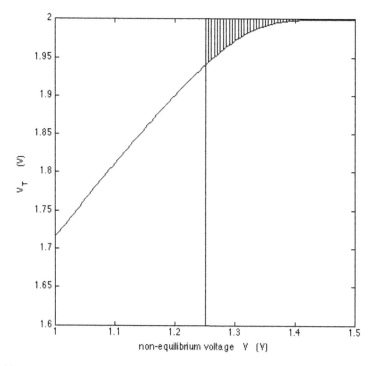

Fig. 3.14 Representation of V_T near the pinch-off voltage, in the moderate and weak inversion regions. The plot is an enlarged view Fig. 3.1. The hatched are represents the drain current divided by β when V_G and V_S are respectively equal to 2 V and 1.25 V (MATLAB fig314.m)

in Fig. 3.14. It represents Eq. 3.6 when ψ_S is evaluated by means of the Charge Sheet Model.

It is obvious that the difference between V_G and V_T does not vanish abruptly but tends to decrease exponentially as predicted by Eq. 2.29. The exponential trend of V_T explains why the current is decreasing exponentially. Four to five 'time constants' is enough to level off the drain current, supporting the earlier made statement that drain-to-source voltages as low as 100 mV suffice to saturate MOS transistors in weak inversion.

3.6 Conclusion

In Chapter 3, a construction allowing to 'visualize' the drain current of CMOS circuits is introduced and a number of examples are reviewed. Although most of the examples relate to strong inversion, the graphical representation applies to all modes of operation. The compact model introduced in the next chapter follows a similar way while making use of an analytical approximation of V_T.

Chapter 4
Compact Modeling

4.1 The Basic Compact Model

Though the C.S.M is very instrumental for understanding the operation modes of MOS transistors, it is not suited for circuit design. More appropriate models have been developed for this purpose, namely the E.K.V. model (for Enz, Krumenacher and Vittoz (Enz and Vittoz 2006)) and the A.C.M. model (for Advanced Compact Model (Cunha et al. 1998)). These belong to a category designated currently by the name of *compact models*. Like the C.S.M, they derive from the gradual channel approximation. More advanced versions encompassing short channel effects and mobility degradation have been developed (Enz and Vittoz 2006), but at the expense of growing complexity. This chapter reviews the basics of the E.K.V and A.C.M models.

What is making the C.S.M. inappropriate for circuit design is the intricacy of the expressions connecting the gate voltage V_G and the surface potential ψ_S to the mobile charge density Q'_i. In the Charge Sheet Model we integrate the right part of Eq. 2.7 with respect to the surface potential. The integration with respect to the mobile carrier density is not considered for an explicit expression of the surface potential versus Q'_i does not exist as reminded by the expression below, which is a replica of Eq. 2.16:

$$d\left(-\frac{Q'_i}{C'_{oc}}\right) = -\left(1 + \frac{\gamma}{2\sqrt{\psi_S}}\right) d\psi_S \qquad (4.1)$$

To get across the difficulty, the E.K.V. and A.C.M. models take advantage of the fact that the term between brackets right varies little, whatsoever the mode of operation, weak or strong inversion, saturation or not. Modern transistors exhibit indeed γ's slightly less than one while the surface potential doesn't vary much. The factor multiplying $d\psi_S$ is assimilated consequently to a constant generally between 1.2 and 1.5, which is called the ***slope factor*** n.[1] The approximation offers the possibility

---

[1] The name 'slope factor' given to n covers slightly different concepts in the literature. In strong inversion, the slope factor is invoked generally in order to model the body effect. In weak inversion, n is given by the maximum of the subthreshold slope.

to integrate Eq. 2.7 with respect to the mobile charge density while getting rid of the surface potential. As a result, Eq. 4.1 is turned into the expression below

$$d\left(-\frac{Q'_i}{C'_{ox}}\right) = -n d\psi_S \tag{4.2}$$

which can be written as follows:

$$d\psi_S = -2U_T dq \tag{4.3}$$

after introducing the **normalized mobile charge density**:

$$q = -\frac{Q'_i}{2nU_T C'_{ox}} \tag{4.4}$$

4.2 The E.K.V. Model

We present hereafter a comprehensive review of the E.K.V. model and stress the fact that the assumption concerning the constant slope factor n paves the way towards analytical expressions of both V_T and I_D. This is the cornerstone of the model.

4.2.1 The $V_T(V)$ Characteristic

We start from Boltzmann statistics like in Chapter 3, taking the total differential of Eq. 3.2:

$$\frac{dQ'_i}{Q'_i} = \frac{dq}{q} = \frac{d\psi_S - dV}{U_T} \tag{4.5}$$

When dq is substituted to $d\psi_S$ according to Eq. 4.3, the above expression becomes a differential equation relating the non-equilibrium voltage to the normalized mobile charge density:

$$-dV = U_T\left(2 + \frac{1}{q}\right) dq \tag{4.6}$$

The integration is performed from source to drain considering respectively q_S and q_D for the normalized charge densities while V_S and V_D represent the non-equilibrium voltage:

$$V_D - V_S = U_T\left[2(q_S - q_D) + \log\left(\frac{q_S}{q_D}\right)\right] \tag{4.7}$$

The scope of this expression is very broad for it is the result of solid-state physics considerations only (Brun et al. 1990). The equation however suffers from a drawback. Saturated transistors give way to very small q_D's making the difference

4.2 The E.K.V. Model

($V_D - V_S$) run out of control. Another presentation would be more appealing; all the more MOS transistors are generally saturated in analog circuits. In the alternative presentation below, the integration limits resume to a constant and a variable q for the mobile charge density. The constant is equal to one and the concomitant voltage *defined* as the **pinch-off** voltage V_P. For the variable mobile charge density, the corresponding limit is non-equilibrium voltages V:

$$V_P - V = U_T \left[2(q-1) + \log(q) \right] \tag{4.8}$$

The normalized mobile charge density q is plotted in Fig. 4.1 versus the difference $V - V_P$. It consists practically of two quasi-linear sections separated by a sharp break when V is nearing V_P.

An upside-down replica of Fig. 4.1 is shown in Fig. 4.2 where the vertical axis has been multiplied by $2nU_T$. The plot represents now Q'_I/C'_{ox}, or the difference $(V_G - V_T)$. The curve is thus an illustration of V_T similar to that of Fig. 3.1, except for the axes; in Fig. 4.2, the threshold voltage is plotted against the gate voltage and the zero of the horizontal axis is the pinch-off voltage, while in Fig. 3.1, V_T is plotted against the substrate.

It is clear that the break in the middle corresponds to what is meant by the pinch-off voltage. The concept entails consequently a clear definition ($q = 1$). What is

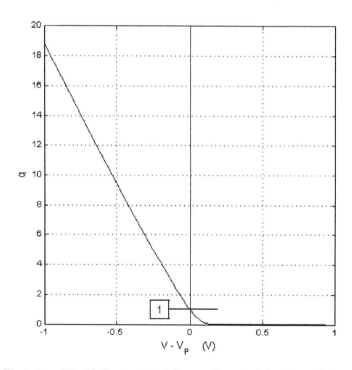

Fig. 4.1 Illustration of Eq. 4.8. The transistor is in strong inversion left and in weak inversion right (MATLAB fig041.m)

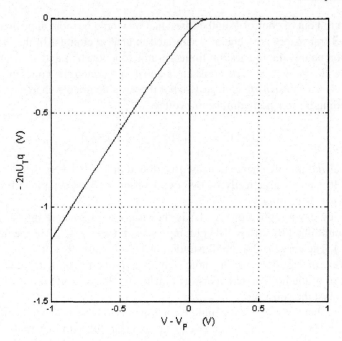

Fig. 4.2 is an upside-down replica of Fig 4.1 after multiplying q by $2nU_T$ so that the vertical axis represents $-Q'_i/C'_{ox}$ or $V_G - V_T$ (MATLAB fig041.m).

still missing however is an expression connecting V_P to the gate voltage V_G. To get this, imagine that q is getting very large so that Eq. 4.8 boils down to:

$$V_P - V = 2qU_T \tag{4.9}$$

After multiplying both sides by the slope factor n, the expression below bridging the pinch-off to the gate voltage is obtained:

$$n(V_P - V) = 2nU_T q = V_G - V_T \tag{4.10}$$

This shows that V_T becomes a linear function of V deep in strong inversion. We may write then:

$$V_T = nV + V^* \tag{4.11}$$

where V^* is a constant that is equal to:

$$V^* = V_G - nV_P \tag{4.12}$$

We can now connect the pinch-off voltage to the gate voltage owing to the fact that the last equation boils down to:

$$V_P = \frac{V_G - V^*}{n} \tag{4.13}$$

4.2 The E.K.V. Model

Only two constants are required thus to relate V_P to V_G and vice-versa, the slope factor n and V^*. We define the latter as the **threshold voltage V_{TO}** of the compact model,[2] turning Eq. 4.13 into:

$$V_p = \frac{V_G - V_{TO}}{n} \qquad (4.14)$$

4.2.2 The Drain Current

For the drain current, we start from Eq. 2.7 reproduced below after replacing the mobile charge density by its normalized counterpart:

$$I_D dx = 2nU_T \mu W [q d\psi_S - U_T dq] \qquad (4.15)$$

Like in the previous section, we take advantage of Eq. 4.3 substituting dq to $d\psi_S$:

$$I_D dx = -2nU_T^2 \mu C'_{ox} W [q + 1] dq \qquad (4.16)$$

The integration with respect to q yields:

$$i = [q^2 + q]_{V_D}^{V_S} \qquad (4.17)$$

after introduction of the **normalized drain current i**:

$$i = \frac{I_D}{I_S} \qquad (4.18)$$

and the **specific current I_S**[3]:

$$I_S = 2nU_T^2 \mu C'_{ox} \frac{W}{L} = 2nU_T^2 \beta \qquad (4.19)$$

Remarkably, the contributions of the drift and diffusion currents are still identifiable for q^2 and q are the counterparts of the Charge Sheet Model corresponding currents. The two equilibrate when q is equal to one. At this point, which corresponds to the pinch-off voltage, the drain current I_D equals twice the specific current I_S.

---

[2] V_{TO} should not be confused with $V_T(0)$. The latter represents the magnitude of V_T when V is equal to and is a function thus of the gate voltage and the pinch-off voltage whereas V_{TO} is a constant.

[3] Slightly different definitions of the specific current are given by Enz and Vittoz (2006) and Cunha et al. (1998). The first makes use of Eq. 4.19 while the second substitutes the factor 0.5 to the factor 2.

The normalized mobile charge or the specific current are currently advocated to differentiate strong from weak inversion. Both measure how deep transistors are in strong (q or $i \gg 1$) or weak inversion (q or $i \ll 1$). The normalized drain current is called therefore also the ***inversion index*** (Enz and Vittoz 2006).

4.2.3 The Equations of the Basic E.K.V. Model

The slope factor n, the specific current I_S and the threshold voltage V_{T0} are the three basic parameters common to the E.K.V. and A.C.M. models. Equation 4.20 reviews the set of equations making up the model. The first line recalls the definition of the normalized drain current. The second line relates the normalized drain current to the normalized mobile charge density and vice-versa (the Charge Sheet Model doesn't have an explicit expression for the second). The third line relates the channel voltage V to the normalized mobile charge density and the pinch-off voltage V_P. Since Eq. 4.20d cannot be inverted, a MATLAB function called ***invq*** is introduced allowing to derive q from V_P and V (see Matlab Toolbox and Annex 2). The fourth line connects the pinch-off voltage to the gate voltage and the threshold voltage V_{T0}.

$$i = \frac{I_D}{I_s} \quad \text{(a)}$$

$$i = q^2 + q \quad \text{(b)} \qquad q = 0.5\left(\sqrt{1+4i} - 1\right) \quad \text{(c)}$$

$$\frac{V_P - V}{U_T} = 2(q-1) + \log(q) \quad \text{(d)} \qquad q = invq\left(\frac{V_P - V}{U_T}\right) \quad \text{(e)}$$

$$V_P = \frac{V_G - V_{T0}}{n} \quad \text{(f)} \tag{4.20}$$

An interesting interpretation can be obtained when the transistor is not saturated (Chatelain 1979). It takes advantage of two normalized drain currents: the ***forward normalized current*** i_F, associated to the source terminal, and the ***reverse normalized current*** i_R, associated to the drain:

$$i_F = q_S^2 + q_S \tag{4.21}$$

$$i_R = q_D^2 + q_D \tag{4.22}$$

With these, the drain current of the non-saturated transistor boils down to the difference of a ***forward*** and a ***reverse*** current each representing the drain current of ***saturated*** MOS transistors whose source voltages are respectively V_S and V_D

$$i = i_F - i_R \tag{4.23}$$

4.2.4 Graphical Interpretation of the E.K.V. Model

The E.K.V. model can be 'visualized' by means of the graphical interpretation presented in Chapter 3. Consider a saturated grounded source transistor whose parameters n, V_{T0} and I_S are respectively equal to 1.2, 0.4 V and 0.7 μA.

The thick dashed line across Fig. 4.3 is a representation of Eq. 4.11, where V_{T0} is put in the place of V^*. We consider three gate voltages respectively equal to 0.60, 0.35 and 0.30 V. The corresponding pinch-off voltages predicted by Eq. 4.20f are marked by circles. To plot the three $V_T(V)$ curves, we proceed by choosing realistic values of q, e.g. a logarithmic scale from 10^{-3} to 10, evaluate V for every V_P by means of Eq. 4.8 and subtract $2nU_Tq$ from the gate voltage like in Fig. 4.2. All curves are copies of a same mold shifted along the dashed line.

We know from Chapter 3, that the hatched areas displayed in Fig. 4.3 stand for the drain currents divided by β and are equal to $2nU_T^2 i$ owing to the definition of I_S given by Eq. 4.19. When the gate voltage is large (0.6 V), the pinch-off voltage is positive while the area representing the drain current has a more or less triangular shape that is typical of strong inversion. As the gate voltage decreases (0.35 and 0.30 V), the pinch-off voltage V_P shifts left becoming negative as the gate voltage

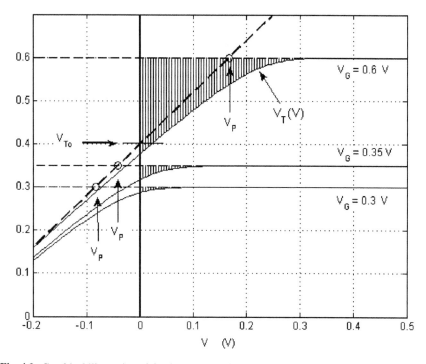

Fig. 4.3 Graphical illustration of the drain current delivered by a saturated grounded source transistor whose gate voltage takes three distinct values. The hatched areas represent the drain currents divided by β (MATLAB fig043.m)

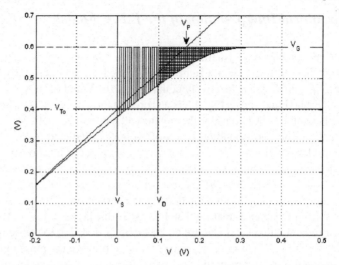

Fig. 4.4 Graphical illustration of the drain current of a non-saturated grounded source transistor. The vertically and horizontally hatched areas stand respectively for the forward and reverse currents. Their difference represents I_D/β (MATLAB fig044.m)

gets smaller than V_{TO}. The area representing the drain current is not only smaller, but it varies exponentially for V is controlled by the $\log(q)$ term, the other term being constant. The drain current varies exponentially with V for we are in weak inversion.

Consider now the same transistor when it is not saturated. The drain voltage is only 0.1 V while the gate voltage is equal to 0.6 V. According to Eq. 4.23, the drain current is represented now by the difference between forward and reverse currents. The vertically and horizontally hatched areas of Fig. 4.4 represent the graphical counterparts of these, respectively $2nU_T^2 i_F$ and $2nU_T^2 i_R$, and the difference the actual drain current.

4.3 The Common Source Characteristics $I_D(V_G)$

To get familiar with the model, we consider a few examples. To begin with, we evaluate the drain current of a grounded source ($V_S = 0$) saturated MOS transistor ($q_R = 0$) whose gate voltage V_G varies from 0 to 1.2 V. The slope factor n is supposed to be equal to 1.2, I_S equal to 0.70 μA and V_{TO} 0.40 V like above. To evaluate the drain current one can proceed along two ways, the 'parametric' or the 'direct' hereafter.

In the parametric method, illustrated by Eq. 4.24, the normalized drain current and the pinch-off voltage are evaluated in terms of an arbitrary q vector taking

4.3 The Common Source Characteristics $I_D(V_G)$

advantage of Eq. 4.20b and d, The drain current I_D and the gate voltage V_G follow with Eq. 4.20a and f.

$$q \Rightarrow \begin{cases} i = q^2 + q & \Rightarrow \quad I_D = i \cdot I_S \\ V_P = U_T (2(q-1) + \log(q)) & \Rightarrow \quad V_G = nV_P + V_{TO} \end{cases} \quad (4.24)$$

In the 'direct' method illustrated by Eq. 4.25, the starting point is the gate voltage V_G, which leads to V_P by means of Eq. 4.20f. The normalized mobile charge density q is extracted from V_P by means of the ***invq*** function of Eq. 4.20e. This leads to the normalized drain current i owing to Eq. 4.20b.

$$V_G \Rightarrow V_P = \frac{V_G - V_{TO}}{n} \Rightarrow q = invq\left(\frac{V_P}{U_T}\right) \Rightarrow i = q^2 + q \Rightarrow I_D = i\, I_S$$
(4.25)

The two methods yield the same results. Figure 4.5 shows the drain current plotted versus the gate voltage and displays the weak and strong inversion approximations of $I_D(V_{GS})$ derived from the EK.V model. These are discussed more in detail in the next section.

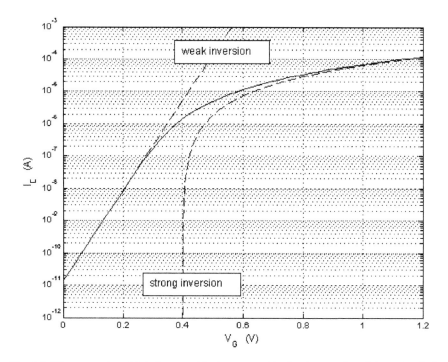

Fig. 4.5 The compact model drain current I_D is compared to the strong and weak inversion drain current approximations (MATLAB fig045.m)

4.4 Strong and Weak Inversion Asymptotic Approximations Derived from the Compact Model

Approximate expressions of the drain current that are valid in weak and strong inversion can be derived easily from the compact model. In strong inversion q is supposed to be large. The pinch-off voltage is thus almost equal to $2U_T q$ and since the drift current overwhelms the diffusion current, q can be neglected with respect to q^2. One has then:

$$V_P = \frac{V_G - V_{TO}}{n} \approx 2U_T q \approx 2U_T \sqrt{i} = 2U_T \sqrt{\frac{I_D}{I_S}}$$

or

$$I_D \approx \mu C'_{ox} \frac{W}{L} \frac{(V_G - V_{TO})^2}{2n}$$

(4.26)

In weak inversion, the opposite holds true. Since q is larger than q^2, the normalized drain current i is equal to q:

$$V_P = \frac{V_G - V_{TO}}{n} \approx U_T(-2 + \log(q)) \approx U_T(-2 + \log(i)) \approx U_T\left(-2 + \log\left(\frac{I_D}{I_S}\right)\right)$$

or :

$$I_D \approx I_S \exp\left(2 - \frac{V_{TO}}{nU_T}\right) \exp\left(\frac{V_G}{nU_T}\right) = I_0 \exp\left(\frac{V_G}{nU_T}\right)$$

(4.27)

The strong and weak inversion approximations of the drain current are illustrated in Fig. 4.5 by means of dashed lines.

4.5 Checking the Compact Model Against the C.S.M.

How to assess the performances of the compact model? What is the impact of the assumption underlying Eq. 4.3? To answer these questions, we compare currents evaluated by means of the compact model to currents predicted by the C.S.M. To do this, we must set up first an acquisition algorithm extracting n, I_S and V_{TO} from C.S.M. currents, second, reconstruct currents by means of the E.K.V model and, third, compare the results to the original data.

4.5.1 The Acquisition Algorithm (MATLAB Identif3.m)

The acquisition algorithm extracting the slope factor, the threshold voltage and the specific current from C.S.M drain currents takes advantage of the common-gate configuration. The configuration is commonly advocated in the literature (Enz and Vittoz 2006; Coltinho et al. 2001). The algorithm that follows proceeds in two steps: first, we evaluate the **unary specific current** I_{Su} (the specific current when W is equal to L), second, the slope factor n and the threshold voltage V_{TO}.

4.5 Checking the Compact Model Against the C.S.M.

We start with I_{Su}. The algorithm takes advantage of the fact that in the pinch-off voltage V_P is constant for it depends only on V_G, which is fixed in the common-gate configuration (see Eq. 4.20f). Changing V_S does not affect the pinch-off voltage thus. The idea is to search the value of I_{Su} that must be plugged in the equations below to keep V_P constant when the source voltage V_S varies modifying the drain current I_{Du}.

$$\frac{I_{Du}}{I_{Su}} = i \Rightarrow q = 0.5\left(\sqrt{1+4i} - 1\right) \Rightarrow V_P - V_S = U_T\left(2(q-1) + \log(q)\right) \tag{4.28}$$

To this effect, we set up a test vector called $I_{Su}{}^*$, which is supposed to encompass the unknown unary specific current I_{Su} usually comprised between 10^{-7} and 10^{-5} A. We consider various source voltages V_S and divide the matching drain currents I_{Du} by the $I_{Su}{}^*$ vector. Then according to the equations above, we evaluate the normalized current vectors i^*, then the normalized mobile carrier density vectors q^* and there from the pinch-off voltage vectors $V_P{}^*$. Because the individual specific currents making out the $I_{Su}{}^*$ vector are all different, the pinch-off voltages listed in every $V_P{}^*$ vector are distinct. All vectors however encompass necessarily the pinch-off voltage V_P. All intersect thus at V_P.

Two source voltages at least are needed, preferably one in strong and one in weak inversion. We consider the ratio R^* of the corresponding $V_P{}^*$ vectors and find I_{Su} when the ratio gets equal to one, which can be done by means of the MATLAB interpolation instruction below:

$$I_{Su} = \mathbf{interp1}\left(R^*, I_{Su}^*, 1, 'cubic'\right) \tag{4.29}$$

We make use of a second interpolation for the pinch-off voltage V_P:

$$V_P = \mathbf{interp1}\left(I_{Su}^*, V_P^*, I_{Su}, 'cubic'\right) \tag{4.30}$$

When more that two source voltages are considered, one gets not one but several unary specific currents. Remarkably, these are practically identical for the differences are generally less than 0.1%.

Now that the unary specific current is known, we can proceed to the second step of the identification algorithm and get n and V_{TO}. The algorithm makes use of the linear dependence of the pinch-off and gate voltages illustrated by Eq. 4.20f. All what is needed thus is to repeat the I_{Su} acquisition algorithm considering not one but several gate voltages and to plot the pinch-off voltages versus the gate voltages. All the points should lie on a straight line whose slope and constant term yield respectively n and V_{TO}. A linear regression takes care of this:

$$\begin{array}{l} P = polyfit(V_P, V_G, 1); \\ n = P(1); \\ V_{TO} = P(2); \end{array} \tag{4.31}$$

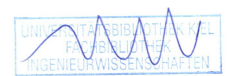

4.5.2 Verification

To assess the correctness of the acquisition algorithm and verify the validity of the model in the same time, we make a test: we set up a number of common-gate C.S.M drain currents, extract the E.K.V parameters and reconstruct the original currents by means of the compact model.

For what concerns the C.S.M, we consider a unary ($W = L$) N-type transistor having a substrate impurity concentration equal to 10^{18} at/cm^3, an oxide thickness of 2 nm and a V_{FB} equal to 0.9 V (V_{FB} controls the threshold voltage but has no impact on the specific current whatsoever). The temperature is 300°K.

We select two source voltages, respectively to 0.1 and 0.6 V, one in weak and one in strong inversion, and consider seven gate-to-substrate voltages from 0.6 to 1.2 V in steps 0.1 V wide. The seven unary specific currents obtained after running the acquisition algorithm display less than 0.1% deviation and yield a I_{Su} of 1.263×10^{-6} A. The slope factor and the threshold voltage, n and V_{To}, are respectively 1.153 and 0.5003.

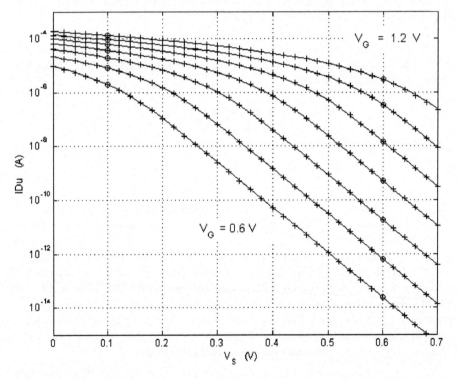

Fig. 4.6 $I_{Du}(V_S)$ characteristics of a saturated common-gate transistor. The continuous lines represent the original currents of the Charge Sheet Model, circles point to the data put to used by the acquisition algorithm and crosses show the reconstructed E.K.V. drain currents (MATLAB fig046.m)

4.5 Checking the Compact Model Against the C.S.M.

Knowing I_{Su}, n and V_{To}, we reconstruct the drain currents by means of the E.K.V model. Figure 4.6 compares the reconstructed to the original C.S.M currents. The continuous lines represent the C.S.M. drain currents, circles mark the strong and weak inversion currents used in order to assess the unary specific currents and crosses represent the reconstructed drain currents.

The fact that the errors are smaller than 1% is clearly the sign that the E.K.V compact model is a good approximation of the Charge Sheet Model, notwithstanding the assumption assimilating the slope factor n to a constant. Whether the parameters are true physical entities is not relevant; all the more that the Charge Sheet Model ignores the concept of threshold voltage. The fact that the model reproduces static drain currents as well as g_m/I_D's over a wide range of terminal voltages with satisfactory accuracy is what matters.

Figure 4.7 compares reconstructed to C.S.M common-source currents considering various back-bias voltages. The correspondence is satisfactory except deep in weak inversion and low back-bias voltages. The explanation is related in all probability to the slope factor decrease in weak inversion illustrated by Eq. 2.38 and the plot of Fig. 2.6. The model does not take this into account.

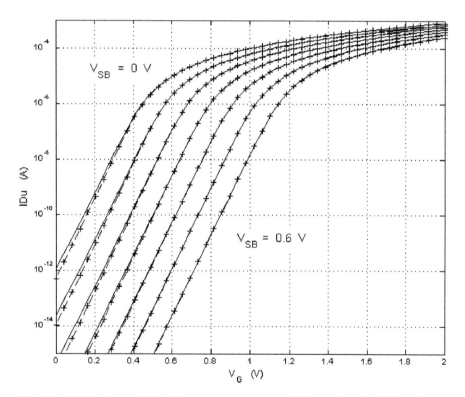

Fig. 4.7 $I_D(V_G)$ characteristics of the same transistor considering various back-bias voltages. The continuous lines represent the drain currents of the Charge Sheet Model. Crosses show the reconstructed drain currents taking advantage of the E.K.V model (MATLAB fig046.m)

4.6 Evaluation of g_m/I_D

An analytical expression of the g_m/I_D ratio in terms of the E.K.V compact model exists contrarily to what happens with the C.S.M. We start from the definition of the transconductance over drain current ratio and take into consideration the fact that the specific current is constant:

$$\frac{g_m}{I_D} = \frac{1}{I_D}\frac{dI_D}{dV_G} = \frac{d\log(I_D)}{dV_G} = \frac{d\log(i)}{dV_G} \quad (4.32)$$

The differentials of $\log(i)$ and V_G are evaluated separately by taking advantage of the expressions listed under Eq. 4.20. We consider moreover a saturated transistor:

$$d\log(i) = \frac{di}{i} = \frac{2q+1}{i}dq$$
and
$$dV_G = n\,dV_P = nU_T\left(2 + \frac{1}{q}\right)dq = nU_T\frac{2q+1}{q}dq \quad (4.33)$$

Hence:

$$\frac{g_m}{I_D} = \frac{1}{nU_T}\frac{q}{i} = \frac{1}{nU_T}\frac{1}{q+1} \quad (4.34)$$

or when q is replaced by i:

$$\frac{g_m}{I_D} = \frac{1}{nU_T}\frac{2}{\sqrt{1+4i}+1} \quad (4.35)$$

In weak inversion, since q and i are much smaller than one, the g_m/I_D ratio is almost constant and equal to $1/(nU_T)$. In strong inversion g_m/I_D declines like the reciprocal of the square root of the normalized drain current. Equation 4.35 leads to an interesting observation moreover: the weak and strong inversion asymptotic approximations of g_m/I_D in a loglog representation cross each other at the point where i is equal to one.

In order to compare g_m/I_D ratios predicted by the compact model and the C.S.M, we consider the same example as above. Since no analytical expression of g_m/I_D ratios is available in the C.S.M, these are evaluated numerically by taking the derivative of the log of the drain current. The results displayed in Fig. 4.8 show that the differences between the C.S.M and compact model representations are almost negligible, except again at very low currents, deep in weak inversion, for the compact model does not take into consideration the slight decrease of the subthreshold slope mentioned earlier.

When plotted versus the drain current instead of the gate voltage, the g_m/I_D of the compact model boils down to a single characteristic, which is the result of the combination of Eqs. 4.18 and 4.35 and reflects the fact that the slope factor and the specific current are supposed to be constants. The plot shown in Fig. 4.9 tends to confirm the observation for all curves tend to merge. The utmost difference, once more, is due to the lessening slope factor in weak inversion. One can summarize by

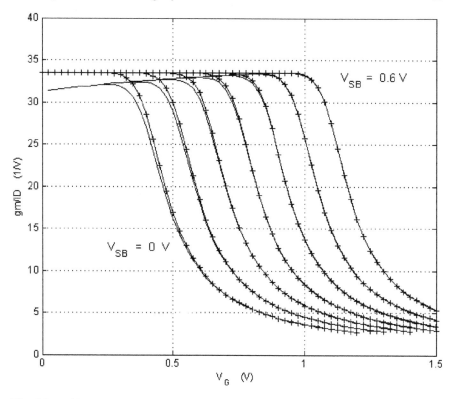

Fig. 4.8 g_m/I_D curves of the same transistor as in Figs. 4.6 and 4.7. *Plain lines* relate to the Charge Sheet Model, crosses to the compact model (MATLAB fig048.m)

saying that this difference is the most prominent effect that results from the basic assumption on which the compact model is based.

4.7 Sizing the Intrinsic Gain Stage by Means of the E.K.V. Model

We derived W/L ratios and drain currents of the Intrinsic Gain Stage in Chapter 1. Only the strong and weak inversion results were demonstrated for an analytic expressions connecting g_m to I_D was lacking. We can now reformulate the problem with the compact model. The starting point is the expression below (see Eq. 1.17) where the numerator g_m is equal to ω_T times and load capacitance C, the g_m/I_D ratio given by Eq. 4.34:

$$I_D = \frac{g_m}{\left(\frac{g_m}{I_D}\right)} = g_m n U_T (1 + q) \tag{4.36}$$

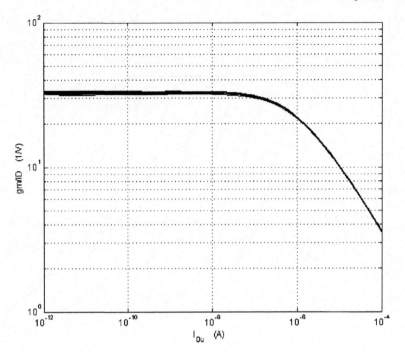

Fig. 4.9 The g_m/I_D curves of Fig. 4.8 plotted against the drain currents. All curves tend to merge, whatsoever the source voltage (MATLAB fig048.m)

Since the factor $g_m n U_T$ represents the minimum drain current I_{Dmin} needed to sustain the gain-bandwidth product ω_T in weak inversion according to Eq. 1.14, the above equation may be rewritten as follows:

$$I_D = I_{D\min}(1+q) \tag{4.37}$$

A second equation is needed in order to connect the aspect ratio W/L to q. This is straightforward for W/L is the ratio of the drain current I_D over the unary drain current I_{Du}, the latter being equal to the specific current I_{Su} times the normalized drain current i, which in turn is the sum of q^2 and q. One has thus:

$$\frac{W}{L} = \frac{I_D}{I_{Su} i} = \frac{I_D}{I_{Su}} \frac{1}{q^2 + q} \tag{4.38}$$

The expression linking W/L to I_D is obtained consequently after eliminating q between Eqs. 4.37 and 4.38:

$$\frac{W}{L} = \frac{I_{D\min}^2}{I_{Su}} \frac{1}{I_D - I_{D\min}} = n \frac{\omega_T^2 C^2}{2K} \frac{1}{I_D - I_{D\min}} \tag{4.39}$$

4.8 The Common-Gate g_{ms}/I_D Ratio

The result is illustrated in Fig. 1.4 by the continuous curve matching the strong and weak inversion asymptotic conducts predicted by Eqs. 1.11 and 1.14.

4.8 The Common-Gate g_{ms}/I_D Ratio

The common-gate g_{ms}/I_D ratio is evaluated like the g_m/I_D ratio, dV_S being substituted to dV_G. To know dV_S we differentiate Eq. 4.20d keeping in mind that V_P is constant:

$$dV_S = -U_T\left(2 + \frac{1}{q}\right)dq = -U_T\frac{2q+1}{q}dq \qquad (4.40)$$

The g_{ms}/I_D is similar to the common-source g_m/I_D, the sign however is opposite while the n factor disappears:

$$\frac{g_{ms}}{I_D} = -\frac{1}{U_T}\frac{2}{\sqrt{1+4i}+1} \qquad (4.41)$$

The g_{ms}/I_D ratio of the model (represented by means of crosses) is compared to its C.S.M. counterpart represented by the continuous line in Fig. 4.10. The gate voltage is equal to 1.2 V.

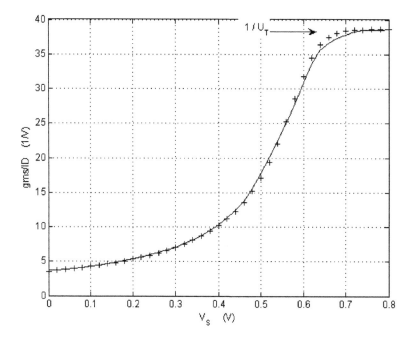

Fig. 4.10 Comparison of g_{ms}/I_D curves obtained with the model (crosses) and Charge Sheet Model (continuous lines) for V_G equal to 1.2 V (MATLAB fig410.m)

4.9 An Earlier Compact Model

The majority of compact models take advantage of the forward and reverse current concept introduced in 1979 by Chatelain (1979). In 1982, Oguey and Cserveny (1982) proposed a continuous model, which turns out to be a mathematical interpolation joining weak and strong inversion approximate equations in a continuous manner. The forward and reverse currents (named respectively F and R as above) are represented by the expression below where V_P and I_S represent the pinch-off voltage and the specific current considered earlier:

$$I_{F,R} = I_S \cdot \log^2\left(1 + \exp\frac{V_P - V_{S,D}}{2U_T}\right) \tag{4.42}$$

When the transistor is saturated, the reverse current is equal to zero for the drain voltage is larger than the pinch-off voltage. The current resumes then to the equation below after replacing V_P by Eq. 4.20f:

$$I_D = I_S \cdot \log^2\left(1 + \exp\left(\frac{V_G - V_{TO}}{2nU_T}\right)\right) \tag{4.43}$$

In strong inversion, where the gate voltage overdrive $V_G - V_{TO}$ is large compared to $2nU_T$, Eq. 4.42 boils down to:

$$I_D = I_S \cdot \left(\frac{V_G - V_{TO}}{2nU_T}\right)^2 = \beta \cdot \frac{(V_G - V_{TO})^2}{2n} \tag{4.44}$$

In weak inversion, the classical exponential approximation is found:

$$I_D = I_S \cdot \left(\exp\left(\frac{V_G - V_{TO}}{2nU_T}\right)\right)^2 = I_S \cdot \exp\left(\frac{V_G - V_{TO}}{nU_T}\right) \tag{4.45}$$

While the asymptotic expressions conform to the strong and weak inversion approximations, what occurs in between is a matter of mathematics, not semiconductor physics. The difference with respect to real drain currents is small, but larger than what can be achieved with the compact model considered throughout this chapter.

A analytical expression of g_m/I_D can be derived also from Eq. 4.43 by taking the derivative of the $\log(I_D)$ with respect to the gate voltage. After lengthy calculations, one has:

$$\frac{g_m}{I_D} = \frac{1}{nU_T} \cdot \frac{1 - \exp\left(-\sqrt{\frac{I_D}{I_S}}\right)}{\sqrt{\frac{I_D}{I_S}}} \tag{4.46}$$

The asymptotic expressions are similar to those predicted by the weak and strong inversion approximations, but not identical. The g_m/I_D ratio is somewhat overestimated in moderate inversion. The approximations in weak and strong inversion cross each other at the point where i is equal to one like in the E.K.V. compact model.

4.10 Modeling Mobility Degradation

The E.K.V. – A.C.M. model like the C.S.M. give a faithful account on the modes of operation of gradual channel MOS transistors, but mobility degradation is ignored. The assumed proportionality between electrical fields and mobile carrier's velocity embodied by Eq. 2.2 holds true only as long as electrical fields do not trespass some limit. Beyond, the rate at which the carrier's velocity increases with the electrical field slows down gradually. When fields are very large, the carriers move almost at constant speed. The phenomenon is designated commonly by the name of *"mobility degradation"*. Short channel MOS transistors are plagued strongly by this phenomenon not only because of their smaller dimensions but also supply voltages not scaling down at the same rate as channel lengths. To contain mobility degradation, modern transistors undergo a series of dedicated implants relaxing the electrical field near the drain.

4.10.1 The Impact of Mobility Degradation on the Drain Current

The dependence of the mobility on the electrical field is a complex matter. Publications deal with the problem (Bücher 1999; Enz and Vittoz 2006). Generally the longitudinal and vertical electrical fields are treated separately and distinct scattering mechanisms invoked. The impact of the longitudinal electrical field on the drain current can be sketched without too much difficulty however. One can make use of the first order approximation below, which has the merit to keep mathematical treatments within acceptable limits:

$$v = \frac{\mu_o}{1 + \frac{\mu_o}{v_{sat}}|E|} \cdot E \qquad (4.47)$$

The factor multiplying the electrical field E is called generally the *'effective mobility'*. When E is small, the effective mobility boils down to the low-field mobility μ_o, and when E is large, mobility declines as the speed of the carriers levels off until it reaches v_{sat}. The low-field mobility μ_o depends on the type of transistor. It is about three times larger for electrons than for holes. The drift saturation velocity of electrons is around $1.53 \times 10^9 \, T^{-0.87}$ cm/s and that of for holes around $1.62 \times 10^8 \, T^{-0.52}$ cm/s (Muller and Kamins 1977). In a loglog scale, the plot representing the velocity versus the electrical field resumes consequently to two lines:

a straight line through the origin for low fields with a slope equal to μ_o and a horizontal line v_{sat} for high fields. The two cross each other at a point called currently the '*critical field*' E_{crit}.

In the Charge Sheet Model, the impact of the longitudinal electrical field on the drain current can be dealt with without too much difficulty thanks to Eq. 4.47, since the mobility is already a function of the integration variable. The diffusion current moreover can be omitted since degradation takes place in strong inversion chiefly. In the compact model, the interpretation is slightly more intricate. The integration of the drift current is supposed to be performed with respect to the normalized mobile charge density q, which is related to the surface potential ψ_S through the approximation given by Eq. 4.3. This changes the electrical field $d\psi_S/dx$ into $-2U_T dq/dx$ so that one has:

$$I_D\, dx = -2n\, U_T^2 \frac{\mu_o C'_{ox}}{1 - 2U_T \frac{\mu_o}{v_{sat}} \frac{dq}{dx}} W\,(2q+1)\,dq \qquad (4.48)$$

The following expression is obtained after rearranging terms:

$$I_D = \left[-2nU_T^2 \mu_o C'_{ox} W\,(2q+1) + 2U_T \frac{\mu_o}{v_{sat}} I_D\right]\cdot \frac{dq}{dx} \qquad (4.49)$$

After integration, one gets the result below where q_S and q_D represent respectively the normalized mobile charge densities at the source and drain as usual:

$$I_D = \left[-I_S\,(q^2 + q) + 2U_T \frac{\mu_o}{v_{sat} L} I_D\, q\right]_{q_S}^{q_D} \qquad (4.50)$$

Equation 4.50 may be rewritten as follows after introduction of the factor θ representing $\mu_o/v_{sat}L$:

$$I_D = I_S \frac{(q_S^2 + q_S) - (q_D^2 + q_D)}{1 + \theta\,(q_S - q_D)} \qquad (4.51)$$

Since the numerator is nothing but the drain current when mobility degradation is ignored, Eq. 4.51 can be rewritten as follows:

$$I_D = \frac{I_{D\ without\ velocity\ saturation}}{1 + \theta\,(q_S - q_D)} \qquad (4.52)$$

This leads to the well-known expression below after $V_D - V_S$ is substituted to the difference of the normalized mobile charge densities. This is acceptable since in strong inversion the difference $q_S - q_D$ is larger than the log term of Eq. 4.7:

$$I_D = \frac{I_{D\ without\ velocity\ saturation}}{1 + \theta_2\,(V_D - V_S)} \qquad (4.53)$$

The plot of Fig. 4.11 compares drain currents with and without mobility degradation considering an N-channel transistor whose V_G is equal to 1 V. The E.K.V.

4.10 Modeling Mobility Degradation

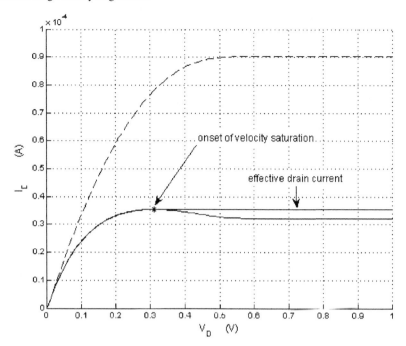

Fig. 4.11 Drain current of a variable mobility transistor (*continuous lines*) compared to the current delivered by the same transistor having a constant mobility (*dashed lines*) (MATLAB fig411.m)

parameters n, I_S and V_{TO} are supposed to be respectively equal to 1.4, 1.2×10^{-6} A. and 0.4 V. The gate length is equal to 100 nm,[4] C'_{ox} is equal to 1.5×10^{-6} F/cm² (tox = 2.3 nm) and v_{sat} equal to 10^7 cm/s. The resultant θ factor is equal to 0.22.

Mobility degradation not only affects the magnitude of the drain current but an unexpected phenomenon is clearly visible just after the maximum. The explanation is the following. As the drain current is nearing its maximum, electrical fields get very large. Since dq/dx varies like the electrical field, the factor between brackets in Eq. 4.49 must get very small in order to keep the drain current constant. When the maximum current is reached, the electrical field is infinite and the expression between brackets equal to zero. This leads to an expression where from we can extract a q_P zeroing the expression between brackets:

$$I_{D\max} = \frac{I_S}{\theta} (2q_P + 1) \qquad (4.54)$$

Beyond the maximum, the sign of the electrical field changes, explaining the decrease of I_D. Drain currents do not decrease actually for the carriers have reached

---

[4] Such short gate lengths require taking into consideration many other short channel effects. The results should be considered as indicative only since many other effects are not considered.

their maximum speed. Since the carrier's density remains unchanged, the drain current comes thus to a horizontal line. Consequently, the point where the drain current is largest is an estimate of the actual pinch-off voltage. This V_P is smaller than the pinch-off voltage of the ideal transistor. We can extract the saturated drain current from Eq. 4.54 if q_p is known. To find this, one must search the q_D zeroing the derivative with respect to q_D of Eq. 4.50. The answer is:

$$q_P = q_S + \left(1 - \sqrt{1 + (2q_S + 1)\theta}\right) \Big/ \theta \tag{4.55}$$

The normalized charge density at the pinch-off point is a function thus of θ and q_S, the latter being a function of the source voltage V_S and the gate voltage V_G through Eq. 4.20d and f. Negative as well as positive values can be found for q_P. Negative q_P's mean that that velocity saturation does not take place yet so that the drain current can still increase. When q_P is positive or equal to zero, velocity saturation is taking place. In the example above, velocity saturation takes place when q_S is equal to 2.13, which yields a V_P of 0.079 V and a V_G equal to 0.51 V. The $I_D(V_D)$ characteristics plotted under Fig. 4.12 show that the onset of velocity saturation starts indeed for gate voltages somewhere between 0.5 and 0.6 V.

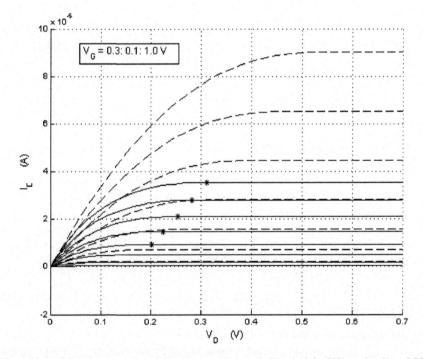

Fig. 4.12 ID(VD) characteristics of the same transistor as in Fig. 4.10 for VG varying from 0.3 V to 1.0 V. Asterisks mark the onset of velocity saturation. Notice the quasi-constant distance separating drain currents in saturation, a feature typical of mobility degradation. The dashed lines represent the drain currents without mobility degradation (MATLAB fig412.m)

4.10 Modeling Mobility Degradation

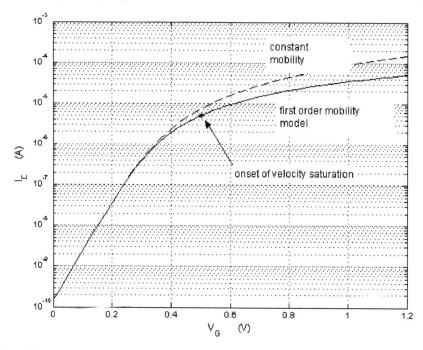

Fig. 4.13 Saturated drain current versus the gate voltage VG of the same transistor as in Fig. 4.12 with and without mobility degradation (MATLAB fig413.m)

Table 4.1 Typical *thetas* for N- and P-channel transistors taking in account the impact of the longitudinal (*theta₁*) and vertical (*theta₂*) electrical fields on the drain current

	$\theta_1(V^{-1})$	$\theta_2(V^{-1})$
NMOS	0.06	0.3
PMOS	0.11	0.14

Figure 4.13 shows the drain current of the saturated transistor versus the gate voltage. Below 0.4 V, mobility degradation doesn't affect the drain current. When the gate voltage increases, the difference with respect to the constant mobility model increases.

Besides the longitudinal electrical field, the vertical field influences strongly also the mobility. Taking this effect into account is more difficult. Often, a second term is added to the denominator of Eq. 4.47 so that the mobility reduction factor takes the form below.

$$\mu = \frac{\mu_o}{1 + \theta_1 (V_G - V_{To}) + \theta_2 (V_D - V_S)} \tag{4.56}$$

Typical values of θ factors are listed in Table 4.1. More elaborated models make use of series expanded in powers of $(V_D - V_S)$ and $(V_G - V_{TO})$.

4.10.2 The Impact of Mobility Degradation on the g_m/I_D Ratio

The impact of mobility degradation on the g_m over I_D ratio of the grounded source transistor is briefly discussed hereafter. The ratio being the derivative of $\log(I_D)$ with respect to V_G, the evaluation proceeds as usual. Two expressions of the g_m/I_D ratio are possible whether velocity saturation takes place or not. In the absence of velocity saturation, g_m/I_D is given by:

$$\frac{g_m}{I_D} = \frac{1}{n\,U_T} \cdot \left(\frac{1}{1+q_S} - \frac{\theta\,q_S}{2\theta\,q_S^2 + (\theta+2)\,q_S + 1} \right) \qquad (4.57)$$

When velocity saturation occurs:

$$\frac{g_m}{I_D} = \frac{1}{n\,U_T} \cdot \frac{2}{(2q_P+1)} \cdot \frac{q_S}{(2q_S+1)} \cdot \left(1 - \frac{1}{\sqrt{1+\theta\,(2q_S+1)}} \right) \qquad (4.58)$$

Figures 4.14 and 4.15 compare the g_m over I_D ratios versus V_G and I_D of the same transistor as above. The impact of mobility degradation is illustrated by the

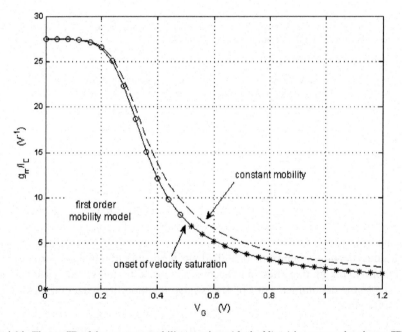

Fig. 4.14 The gm/ID of the constant mobility transistor (*dashed lines*) is compared to the gm/ID of the same transistor making use of the first order mobility model. The continuous lines are obtained by taking the numerical derivative of the log of the current displayed in Fig. 4.13. The circles and asterisks represent gm/ID evaluated respectively by means of Eqs. 4.57 and 4.58 (MATLAB fig414.m).

4.10 Modeling Mobility Degradation

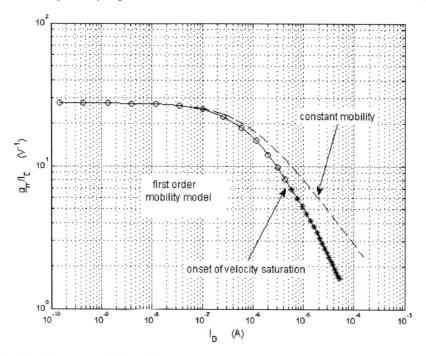

Fig. 4.15 Same data as in Fig. 4.14 versus the drain current (MATLAB fig414.m)

difference between the two curve, the upper curve relating to the ideal transistor, the lower curve to mobility degradation. Circles correspond to Eq. 4.57, asterisks to Eq. 4.58.

Figure 4.15 shows a representation of g_m/I_D versus the drain current like in Fig. 4.9. It is clear that the impact of mobility degradation gets serious in strong inversion. This makes questionable attempts to identify the specific current from the intersection of weak and strong inversion asymptotes.

4.10.3 Sizing the Intrinsic Gain Stage in the Presence of Mobility Degradation

Mobility degradation requires to enlarge W/L's for more current is needed in order to compensate the loss of transconductance in strong inversion. Figure 4.16 shows the impact of the longitudinal electrical field on the Intrinsic Gain Stage. The transistor is the same as above while the g_m/I_D is given by Eq. 4.57 and 4.58. Notice that the result is still too optimistic for the influence of the vertical electrical field is not taken into consideration.

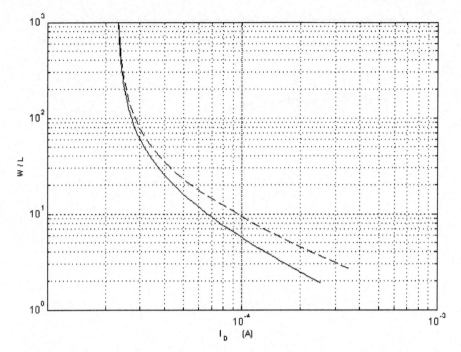

Fig. 4.16 Larger W/L ratios are needed in order to counteract mobility degradation due to the longitudinal electrical field

4.11 Conclusion

The E.K.V.–A.C.M. compact model is a straightforward and rather accurate substitute to the Charge Sheet Model of Chapters 2 and 3. It relies on an approximation that allows replacing the differential of the surface potential by the differential of the mobile carrier's density. This simplifies considerably the computations. Only three parameters are required, the subthreshold factor n, the specific current I_S and the threshold voltage V_{TO}. The model is continuous from weak to strong inversion, but ignores the impact of mobility degradation and the influence other effects inherent to short channel devices. It does not provide a direct connection moreover to physical parameters like the oxide thickness, substrate doping, and temperature what the Charge Sheet Model does.

In the next chapter, we are going to show that the introduction of variable parameters embodies the simple E.K.V. model with the faculty to predict drain currents and g_m/I_D and g_d/I_D ratios of real transistors with satisfactory accuracy, even with short channel devices. The price to pay is the introduction of look-up tables, but the availability of few analytical expressions is an undeniable asset as far as sizing.

Chapter 5
The Real Transistor

The basic E.K.V. model considered in the previous chapter is not suited for real transistors for it makes use of the "gradual channel" approximation, like the C.S.M. Non-uniform doping, mobility degradation, short channel effects, etc. are ignored. Advanced models like BSIM and PSP, which are primarily circuit simulation tools, take care of these but don't offer the degree of flexibility that is desirable.

We show in this chapter that as long as the source and drain voltages with respect to the substrate remain constant, DC currents, g_m/I_D and g_d/I_D ratios of real transistors, even sub-micron devices, can be reconstructed by means of the basic E.K.V model. Once V_S or V_D is modified, the parameters must be updated. The model remains unchanged however.

5.1 Short Channel Effects

Figure 5.1 shows the constituents of a short channel transistor. The region under the thin oxide in the middle embodies the active region. The rest makes up passive parts. The source and drain consist of narrow n- implanted regions, which run on larger n + diffused regions, themselves connected to the contact regions through silicide layers.

With long channel devices, the proportion of fixed charge below the inversion layer that is controlled by the gate is always much larger than that controlled by the source and drain. Consequently, as the gate length decreases the threshold voltage remains practically constant for the gate-controlled charge varies almost like the gate length. This isn't true with short channel devices. The depleted charge controlled by the gate decreases faster than the gate length owing to the consistent contributions of the source and drain. As a result, the threshold voltage begins to roll-off. Dedicated ion implantations help postponing the effect but generally at the expense of a slight increase of the threshold voltage just before roll-off, an effect called the ***reverse short channel effect***.

Short channel roll-off is not the only issue. The drain voltage influences also the threshold voltage. As the drain voltage increases, the drain takes over a larger share of the depleted region previously controlled by the gate. The threshold voltage

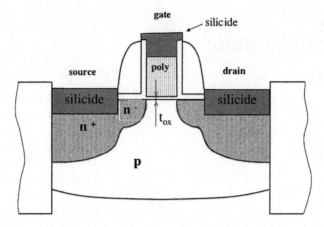

Fig. 5.1 A typical short channel N-MOS transistor

decreases thus, an effect designated currently by the name ***Drain Induced Barrier Lowering*** (**D.I.B.L.**). Contrarily to roll-off, D.I.B.L. is bias dependent and therefore a source of non-linear distortion for it affects V_{T_0} dynamically.

Other effects plague short channel devices, like mobility degradation caused by high electrical fields. With short channel devices, the electrical field along the channel increases rapidly not only for gate lengths are getting smaller but also because supply voltages often don't scale down at the same rate. The increasing mobility degradation produces a loss of current capability that can be partly compensated by the introduction of lowly doped stripes bridging the channel to the drain region as illustrated in Fig. 5.1. These lessen the impact of the longitudinal electrical field somehow but don't restore the original transconductances.

The model considered in the previous chapter doesn't take any of these effects in consideration. However, real $I_D(V_{GS})$ characteristics look very similar to characteristics predicted by the Charge Sheet and E.K.V.–A.C.M. models. A quasi-exponential region and a more or less quadratic strong inversion region are clearly identifiable. Does this mean that it is possible to reconstruct $I_D(V_G)$ characteristics with the compact model of Chapter 4 even with short channel devices? As long as the source-to-substrate and the drain-to-substrate voltages don't change, the answer may be yes. The spatial distributions of the charge in the inversion layer and in the depleted region underneath may explain this. The thickness of the inversion layer is small compared to the gate length (even down to 100 nm gate lengths). To push a little, what happens in the inversion layer boils down to a 1D problem while in the depleted region beneath, things are different. As 'long channel' conditions prevail more or less in the inversion layer, the spatial distribution of the electric field in the depleted region conforms to a 2D problem owing to the large source and drain contributions. This may be the reason why $I_D(V_G)$ characteristics of submicron transistors can be modeled reasonably well with the compact model as long as the source and drain voltages don't change, which implies that all the parameters must be updated as soon as one of these is modified.

5.2 Checking the Validity of the Compact Model when its Parameters vary with the Source and Drain Voltages

The sample displayed in Fig. 5.2[1] represents drain currents of a 10 μm wide N-channel MOS transistor whose drain-to-source voltage is stepped from 0.2 to 1.2 V, considering two source voltages V_S (0 and 0.8 V) and two gate lengths (0.1 and 1 μm). Although distinct, all curves are similar.

The question is: can we reconstruct each of these characteristics dependably by means of the compact model of the previous chapter taking advantage of parameters that depend on the source and drain voltages? To answer the question, we must compare drain currents predicted by the model to real $I_D(V_{GS})$ characteristics. An identification algorithm is needed up therefore.

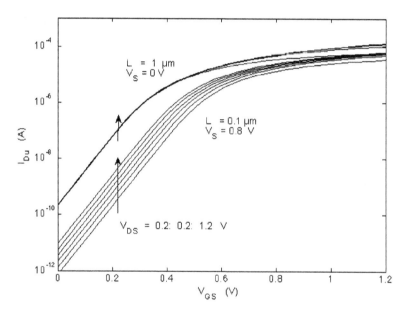

Fig. 5.2 Drain currents of an N-channel unary transistor (W/L = 1) considering two gate length 0.1 and 1 μm, various drain-to-source and back-bias voltages. The device belongs to a low-power, low-voltage 90 nm technology developed by IMEC (Courtesy of IMEC)

---

[1] The currents shown in this figure are reconstructed drain currents obtained by means of the PSP compact MOSFET model. The parameters were extracted from measurements carried out on real physical transistors (courtesy of IMEC). The assumption that reconstructed currents agree fairly well with the physical currents is accepted implicitly. The PSP compact MOSFET model is a product of Philips Semiconductors and Penn State University (now respectively NXP and Arizona State University) (PSP 2006).

5.2.1 E.K.V Parameter Identification (MATLAB IdentifDemo.m)

The identification algorithm[2] makes use of the E.K.V equations introduced in the previous chapter. These are divided in two groups, general equations:

$$V_P - V_S = U_T \left(2\left(q_F - 1\right) + \log\left(q_F\right)\right) \tag{5.1}$$

$$V_P - V_D = U_T \left(2\left(q_R - 1\right) + \log\left(q_R\right)\right) \tag{5.2}$$

$$i = q_F^2 + q_F - q_R^2 - q_R \tag{5.3}$$

and equations involving the parameters n, V_{T0} and I_S:

$$V_P = \frac{V_G - V_{T0}}{n} \tag{5.4}$$

$$I_D = i I_S \tag{5.5}$$

Before we review the acquisition algorithm, three preliminary remarks ought to be made. The first concerns the transistor configuration for data acquisition. The algorithm described in the previous chapter cannot be used for the acquisition method makes use of the common-gate configuration, violating thus the conditions formulated above regarding constant source and drain voltages. The parameters must be extracted from $I_D(V_{GS})$ characteristics exclusively.

The second remark concerns the reference terminal when carrying out measurements. The reference terminal is generally the source of the transistor while the substrate is back-biased. This requires to rewrite the equations above accordingly. The pinch-off voltage of Eq. 5.1 becomes V_{PS}, the left part of Eq. 5.2 is replaced by $V_{PS} - V_{DS}$ while the expression below is substituted to Eq. 5.4. Notice that the threshold voltage V_{T_o} (with a lower case zero) below is defined also with respect to the source.

$$V_{PS} = \frac{V_{GS} + V_{T_o}}{n} \tag{5.6}$$

The third remark concerns geometry. All the 'experimental' drain currents are divided by W/L prior to identification. We consider only **unary drain currents** $I_{Du}(V_{GS})$ and **unary specific currents**.

Let us consider now the acquisition algorithm. The two parameters that are identified first are the slope factor and the threshold voltage. Both emanate from the derivative of the $\log(I_D(V_{GS}))$ characteristics, in other words from g_m/I_D. The slope factor is derived from the maximum of g_m/I_D as usual, while the threshold

---

[2] The identification algorithm can be found in the 0start directory under IdentN.m and IdentP.m. The algorithm makes use of the 'semi-empirical' N- and P-channel data listed under n90.mat and p90.mat. The compact model parameters outputted by the identification algorithm are stored under ParamN.mat and ParamP.mat,. These are turned into **global variables** when running Glob.m (see also Annex 1).

5.2 Checking the Validity of the Compact Model

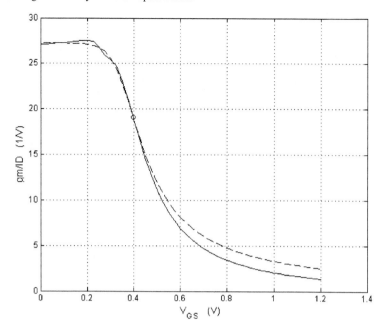

Fig. 5.3 Reconstructed and original g_m/I_D's (respectively dashed and plain lines) of a grounded source N-channel transistor whose L is equal to 100 nm and V_{DS} equal to 0.6 V. The circle corresponds to R equal to 0.7. The MATLAB IdentifDemo.m file illustrates the identification mechanism when the variable M on top of the data list is made equal to one[3]

voltage is the result of a fitting procedure illustrated by Fig. 5.3. The idea is to search the V_{To} that forces the g_m/I_D predicted by the E.K.V model to pass through a predefined point $(g_m/I_D)_o$ of the 'experimental' g_m/I_D curve supposed to lie in moderate inversion, say 80% to 50% below the maximum of g_m/I_D. The acquisition starts with the evaluation of the normalized mobile charge density q_{Fo}, which is derived from the ratio $(g_m/I_D)_o$ over the maximum g_m/I_D, called R, and Eq. 4.34:

$$R = n\, U_T \left(\frac{g_m}{I_D}\right)_o = \frac{1}{1 + q_{Fo}} \qquad (5.7)$$

Knowing q_{Fo}, we evaluate the pinch-off voltage V_{PSo} by means of Eq. 5.1:

$$V_{PSo} = U_T \left(2\left(q_{Fo} - 1\right) + \log\left(q_{Fo}\right)\right) \qquad (5.8)$$

This allows extracting the threshold voltage from the expression below derived from Eq. 5.6, where V_{GSo} represents the gate-to-source voltage at the selected coincidence point:

$$V_{To} = -nV_{PSo} + V_{GSo} \qquad (5.9)$$

---

[3] The MATLAB file IdentifDemo.m illustrates dynamically the evolution of Fig. 5.3 when the drain voltage symbolized by a vertical landmark is swept from 0 to 1.2 V.

The extraction method is fairly reliable for changes of R by 5–10 cent do not affect V_{To} by more than 1–2 mV.

Now that n and V_{To} are known, we identify the unary specific current. The evaluation is straightforward. All what is needed therefore indeed is to divide the drain current I_{Duo} (the drain current at the point considered for the evaluation of V_{To}) by the normalized drain current i_o, which is know since q_{Fo} has been assessed already.

In a nutshell, the slope factor is extracted from the subthreshold drain current characteristic, the threshold voltage from the progressive bending of the drain current in moderate inversion and the specific current from the drain currents coincidence.

One may argue that Eq. 5.7 supposes that the transistor be saturated, which may not be the case. To take care of non-saturation, the expression below, demonstrated further under Eq. 5.20, must substituted to Eq. 5.7:

$$R = \frac{1}{1 + q_{Fo} + q_{Ro}} \quad (5.10)$$

The introduction of q_{Ro} requires however having at one's disposal an additional expression linking q_{Ro} to V_{DS}. To get this equation, we subtract Eq. 5.2 from Eq. 5.1:

$$V_{DS} = U_T \left(2(q_{Fo} - q_{Ro}) + \log\left(\frac{q_{Fo}}{q_{Ro}}\right) \right) \quad (5.11)$$

Equations 5.10 and 5.11 form a system of non-linear implicit equations that can be solved by means of MATLAB interpolation instructions. All what is needed therefore is to generate a *logspace* vector q_R encompassing all possible reverse normalized mobile charge densities and extract the corresponding forward q_F's from Eq. 5.10. One makes then use of Eq. 5.11 to find the concomitant drain to source voltage vector U_{DS}. The q_{Fo} to be put in Eq. 5.8 is found by running the MATLAB interpolation instruction below making use of the U_{DS} and q_F vectors. Notice that the problem requires to be solved only once.

$$q_{Fo} = \mathbf{\textit{interp1}}\,(U_{DS}, q_F, V_{DS}, \text{'cubic'}) \quad (5.12)$$

The reconstructed g_m/I_D curve (represented by the dashed line in Fig. 5.3) calls for a few comments. The 'experimental' and reconstructed curves coincide of course at the reference point. Differences appear else. In weak inversion, the model operates like a filter wiping out a number of local disparities that may be inherent to the 'experimental' data or reflect physical effects like side currents or a shift of the drain current in volume (with P-channel transistors namely). The fact that these differences are smeared out does not represent a problem per see but raises the question as how to define the maximum of g_m/I_D and, more specifically, what is the impact of small variations of the slope factor n on the final threshold voltage V_{To}? The answer is little, for small variations of n are synonymous of small variations of R and the threshold voltage does not depend much on R. The departure in strong inversion between original and reconstructed g_m/I_D curves is more serious. This difference is discussed in the next section.

A final remark concerns the source voltage of the reverse transistor. Since it is not the same as that of the original transistor, results may be questionable. Yet, examples show that Eq. 5.12 yields generally more consistent results than Eq. 5.7.

5.2.2 How to Introduce Mobility Degradation?

After g_m/I_D, let us reconstruct the drain current. The result is shown in Fig. 5.4. In weak and moderate inversion the 'experimental' and model-driven curves coincide practically, but diverge substantially in strong inversion. The reason is that we didn't take mobility degradation into consideration. In the acquisition method described in the previous section, the mobility factor μ that appears in the specific current expression of Eq. 4.19 is evaluated at the reference point, in other words in moderate inversion. It is supposed not to vary. Mobility degradation can be modeled however by making μ a function of the electrical field like in Chapter 2.

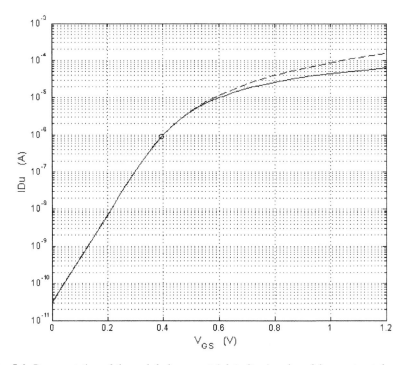

Fig. 5.4 Representation of the real drain current (*plain lines*) and model-reconstructed current (*dashed lines*) making use of the three parameters identified so far. The circle refers to the point selected for the threshold identification (see curve M = 2 of the MATLAB IdentifDemo.m file)

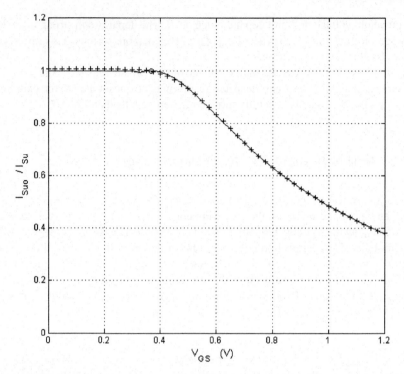

Fig. 5.5 The impact of mobility degradation in strong inversion is illustrated by the roll-off of the plain line curve representing the ratio of the constant weak inversion specific current I_{Suo} over the effective specific current I_{Su}. Crosses represent the approximation based on the theta polynomial $\theta(i)$. The gate length is 100 nm, the source is grounded and the drain voltage equal to 0.6 V like in the two previous figures (see curve M = 3 of the MATLAB IdentifDemo.m file)

The continuous curve of Fig. 5.5 shows the ratio of the 'semi-empirical' over model-reconstructed current represented by means of a dashed line in Fig. 5.4. Two regions are clearly identifiable. Left, the ratio is more or less constant and equal to one for the reconstructed and 'experimental' drain currents coincide practically. Above 0.4 V, mobility degradation is taking over steadily. One can model the trend by turning the unary specific current into a variable. We can define I_{Su} for instance as the product of the constant specific current I_{Suo} of the previous section times a function that rolls-off progressively in strong inversion. In weak inversion, the specific current I_{Su} boils down to the constant **weak inversion unary specific current** I_{Suo}. Else, mobility degradation is acknowledged by dividing μ by a polynomial function (like in Eq. 4.56). Generally, the polynomial is expanded versus V_{GS} and V_{DS}. We take a different approach. We expand the polynomial versus the normalized drain current instead of the gate and drain voltages. The idea is to coalesce the effects of the gate, drain and source voltages by means of the sole normalized drain current. The plot illustrated by crosses in Fig. 5.5, which makes use of a fourth order polynomial fit $\theta(i)$, shows that this is feasible. Eventually a third or even second order polynomials can be put to use.

5.2.3 Drain Current Reconstruction

In the previous section, we described the acquisition of the parameters, n, V_{To}, I_{Suo} and the fitting polynomial $theta(i)$.[4] Figure 5.6 compares drain currents predicted by the model (dots) to the data shown in Fig. 5.2. The two match reasonably well. Relative errors are of the order of 1–2% in moderate and strong inversion. In weak inversion, they attain 4–8% because the model doesn't take into consideration the gradual decline of the g_m/I_D ratio mentioned before. The errors with P-channel transistors are generally larger reaching eventually 10–15% over 8 decades drain current. The reason is probably due to the different nature of the inversion layer, which may be deeper in the substrate than with N-channel transistors.

A 3D representation comparing experimental and reconstructed drain currents versus the drain and gate voltages is represented in Fig. 5.7.

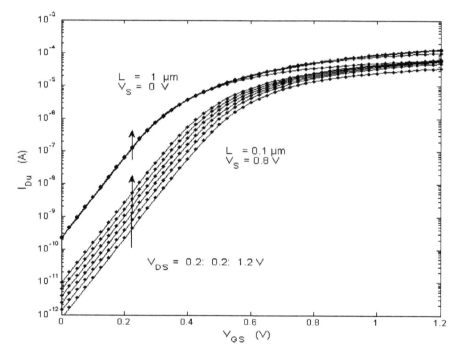

Fig. 5.6 This figure compares reconstructed drain currents (*dots*) to the currents of Fig. 5.2 (*plain lines*) where from the E.K.V. parameters were evaluated by means of the identification algorithm described in the previous section

---

[4] The MATLAB IdentifN.m and IdentifP.m files implementing the acquisition algorithm can be found in the Glob directory together with the 'semi-empirical' data where from the compact model parameters are extracted. It is possible to retrieve the extraction algorithm with other 'experimental' data when desired. To get familiar with the data organization, please consult Annex 1.

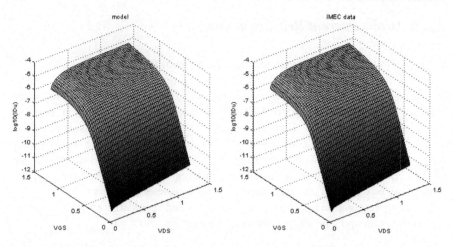

Fig. 5.7 3D representations of the drain current of the 100 nm N-channel grounded source transistor. The reconstructed model is shown *left*, the original *right*. The gate voltage varies from 0 to 1.2 V and the drain from 0.15 to 1.2V

5.3 Compact Model Parameters Versus Bias and Gate Length

The fact that the model reproduces characteristics of short channel devices with few parameters opens interesting prospects. Measurements carried out on large numbers of 'identical' transistors pave the road towards sensitivity analyses by assessing mismatches affecting the slope factor, threshold voltage and specific current. The impact of the temperature can be transposed likewise in terms of parameters sensitivities (see Annex 3). Small modifications of the terminal voltages can be expressed in terms of parameter modifications moreover, which allow evaluating small signal parameters. In a nutshell, the possibility to scrutinize the dependence of the parameters on the gate length and bias conditions opens interesting investigation fields. A few examples are reviewed hereafter.

5.3.1 The Influence of the Gate Length on the Model Parameters

The gate length brings to the fore a number of well-known effects, such as threshold voltage roll-off, reverse short channel effect, D.I.B.L. and C.M.L.

The plot of Fig. 5.8 illustrates the impact of the gate length on the slope factors of N- and P-channel transistors. The slope factors tend to increase when the gate length is shrinking. The effect is more pronounced for P- than for N-channel devices owing to their distinct structure. The drain voltage has very little effect on the slope factor.

The curves of Fig. 5.9 illustrate the influence of the gate length on V_{T0}. The threshold voltage of long channel devices does not depend practically on the gate length nor the drain voltage, whether N or P channel transistors are considered.

5.3 Compact Model Parameters Versus Bias and Gate Length

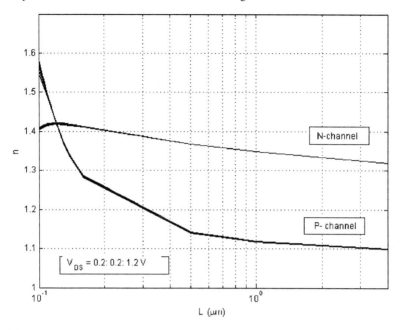

Fig. 5.8 Plot of the subthreshold slope n versus the gate length of grounded source N and P channel transistors for six equally spaced drain voltages (MATLAB SlopeFact1.m)

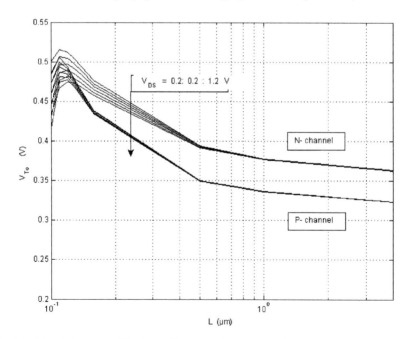

Fig. 5.9 Plot of the threshold voltage V_{To} versus the gate length of grounded source N and P channel transistors considering six equally spaced drain voltages comprised between 0.2 and 1.2 V (MATLAB ThresVolt1.m)

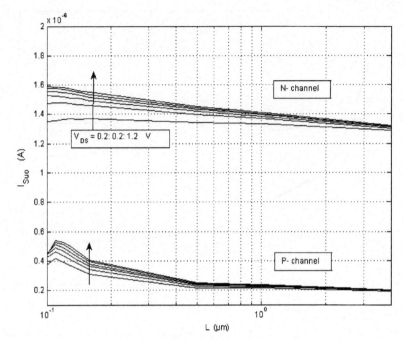

Fig. 5.10 Plot of the weak inversion specific current I_{Suo} versus the gate length of the grounded source N and P channel transistors considering six equally spaced drain voltages comprised between 0.2 and 1.2 V (MATLAB SpecCur1.m)

Below 1 μm, the threshold voltage starts to increase progressively until a rapid roll-off occurs at short gate lengths. The global increase, called the **reverse short channel effect**, reflects the actions taken during fabrication in order to postpone roll-off. The rise contrasts sharply with the abrupt **roll-off** due to the source and the drain depleted regions taking over a larger share of the gate-controlled depleted region. It shows that one is getting close to the minimum achievable gate length.

The data displayed in Fig. 5.10 illustrate the influence of the gate length on I_{Suo}. Though the W/L ratio of unary transistors is constant and equal to one, unary specific currents increase slightly when the drain voltage increases. The widening depleted region near the drain is shortening indeed the effective gate length. As a result, I_{Suo} tends to increase. The effect is commonly designated by the acronym C.L.M for **Channel Length Modulation**.

5.3.2 The Influence of Bias Conditions on the Parameters

The next figures illustrate the influence of the drain-to-source and source-to-substrate voltages. The impact of the drain-to-source voltage on the slope factor n is relatively small and can be ignored as shown already in Fig. 5.8. The influence

5.3 Compact Model Parameters Versus Bias and Gate Length

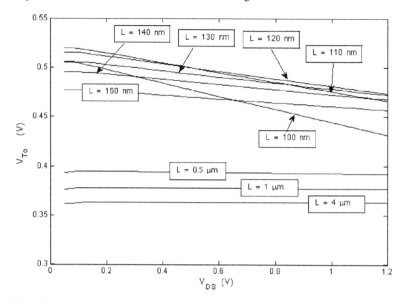

Fig. 5.11 The threshold voltage of the N-channel transistor exhibits almost a linear dependence on the drain-to-source voltage over a wide range, whatsoever the gate length (V_S is equal to zero) (MATLAB ThresVolt2.m)

of V_{DS} on the threshold voltage is more central as shown in Fig. 5.11. As V_{DS} increases, the drain takes over a larger share of the control exercised by the gate, especially when the gate length is small. This lowers the potential barrier carriers must overcome to reach the drain. As a result, the threshold voltage decreases. The effect is designated by the acronym D.I.B.L *for Drain Induced Barrier Lowering*. It affects strongly the derivative dV_{To}/dV_{DS}, called the sensitivity factor S_{VTo}, that characterizes the quasi-linear evolution of V_{To}. S_{VTo} is of the order of -0.13 mV/V with the 4 μm transistor but reaches -66 mV/V and -85 mV/V with the 100 nm transistor considering back-bias voltages respectively equal to 0 and 0.8 V. While negligible when L is larger than 2 μm, D.I.B.L plays a major role with submicron transistors.

A number of other effects are visible also in the same figure. The global shift upwards with shorter gate lengths illustrates the reverse short channel effect mentioned in connection with Fig. 5.9. Threshold voltages grow until the trend changes once roll-off starts to take place below 130 nm. The P-channel transistor is a little less sensitive to D.I.B.L. Its S_{VTo} is equal to -34 mV/V for 100 nm transistors and vanishes faster than with N-channel transistors.

Figure 5.12 displays the influence of back-bias on the threshold voltage of the 100 nm N- and P-channel transistors considering several drain-to-source voltages. The systematic increase of the threshold voltage is an illustration of the well-known **body effect**. It is more pronounced for the N- than for the P-channel transistors.

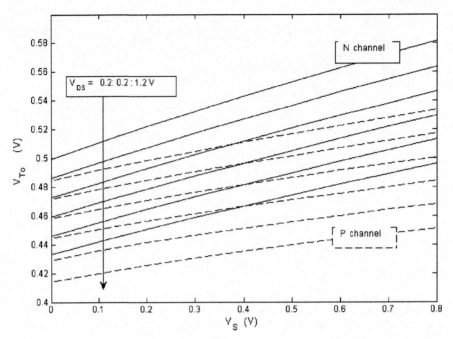

Fig. 5.12 Plot of the threshold voltage V_{To} versus the source-to-substrate voltage V_S considering N- and P-channel transistors with a gate length of 100 nm and six equally spaced drain-to-source voltages (MATLAB ThresVolt3.m)

The impact V_{DS} and V_{GS} have on the specific current I_{Su} is trickier but exemplifies several interesting effects. The overall decline of I_{Su} that is clearly visible in the 3D representation of Fig. 5.13 when the gate-to-source voltage trespasses 0.4 V reflects the growing mobility degradation caused by the electrical field. Below 0.4 V, V_{GS} has little effect but the impact of the drain-to-source voltage is subtler. When V_{DS} decreases, the longitudinal field lessens so that the specific current should be increasing instead of decreasing. Mobility degradation is not the only item to consider however for the specific current depends also on C.L.M. When the drain voltage decreases, the channel length increases slightly owing to the lessening W/L ratio. Mobility and C.L.M impact the specific current in opposite directions thus. The first tends to decrease, the second to increase I_{Su}. According to Fig. 5.13, C.M.L overwhelms mobility degradation in weak and moderate inversion. In strong inversion, the explanation is a bit trickier and requires separating more clearly the impact of D.I.B.L and C.M.L. Fig. 5.14 proposes an interpretation.

The figure represents the unary specific current I_{Su} divided by I_{Suo}, in other words the reciprocal of '*theta*' function. Dividing the specific current by I_{Suo} eludes C.L.M. The fact that the ratio remains practically equal to one in weak and partly in moderate inversion whichever V_{DS} supports the idea. When the gate-to-source voltage trespasses 0.4–0.5 V and mobility degradation starts to grow, we observe a smooth lift up in the non-saturated region. In this region, the longitudinal electrical field

5.3 Compact Model Parameters Versus Bias and Gate Length

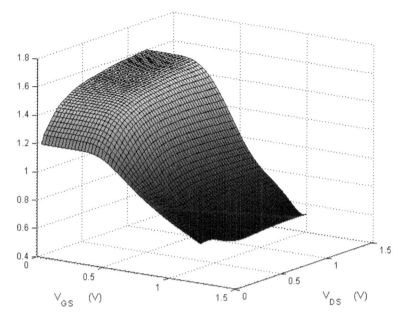

Fig. 5.13 Illustration of the dependence of I_{Su} on the gate and drain voltages for the grounded-source N-channel 100 nm transistor (MATLAB SpecCur2.m)

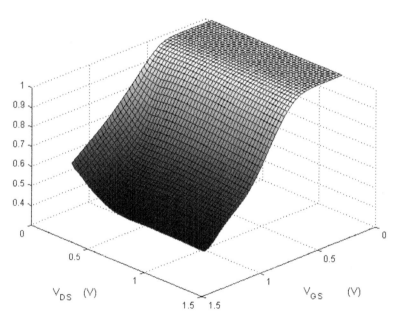

Fig. 5.14 3D representation of the reciprocal of the theta polynomial considering the 100 nm N-channel transistor with zero back-bias (MATLAB SpecCur2.m)

is lessening, mobility degradation decreases thus. Ultimately, when V_{DS} is equal to zero, only the vertical electrical field remains. It looks like if the '*theta*' function discriminates the contributions of the vertical and longitudinal electrical fields. The interpretation of Fig. 5.13 is more intricate for mobility degradation and C.L.M combine their effects with non-saturation.

5.4 Reconstructing $I_D(V_{DS})$ Characteristic

The crucial role played by bias dependent parameters is clearly illustrated when we reconstruct $I_D(V_{DS})$ characteristics. We proceed like in the previous chapter. The specific current is multiplied by the normalized drain current, which requires to know the normalized mobile charge densities q_F and q_R, themselves derived from the applied voltages and the pinch-off voltage. But, contrarily to what happens in the Charge Sheet model where the drain current remains practically constant in saturation, in the compact model the current varies for all the parameters vary with V_{DS}. With short channel devices, the forward mobile carrier density increases with V_{DS} for the threshold voltage decreases owing to D.I.B.L, whereas in long channel devices the gate length decreases owing to C.L.M.

Figures 5.15 and 5.16 compare reconstructed drain currents (represented by means of crosses) to original (continuous) drain currents considering the same

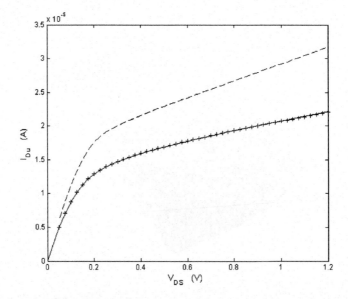

Fig. 5.15 Comparison of the drain currents of the 100 nm grounded source N- channel transistor when V_{GS} is equal to 0.70 V. The *plain line curve* represents the 'experimental' data, crosses illustrate the predicted drain current. The *dashed lines* relate to the model when the mobility is supposed to be invariant (MATLAB fig515.m)

5.4 Reconstructing $I_D(V_{DS})$ Characteristic

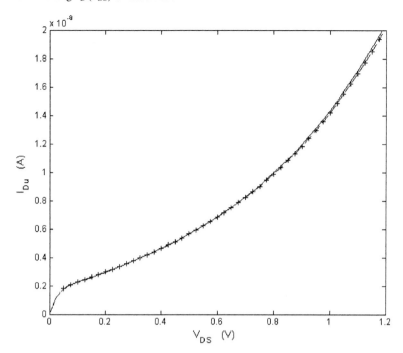

Fig. 5.16 Comparison of the drain currents of the 100 nm grounded source N-channel transistor when V_{GS} is equal to 0.20 V (weak inversion). The *plain line curve* represents the 'experimental' data, crosses the predicted drain current. There is no difference between the *dashed* characteristic (constant mobility) and the current predicted by the model (MATLAB fig515.m)

100 nm N-channel grounded source transistor as above. Two distinct gate voltages, are contemplated, respectively 0.70 and 0.20 V. With the first, the transistor is in strong inversion, with the second it is in weak inversion. Notice that the dashed curve in the first figure represents the drain current without mobility degradation. The curve lies definitely above the actual drain current while in the second figure the curves coincide for mobility does not take place.

Predicted and 'experimental' drain currents differ by less than a few per-cent. The large dissimilarity between the two figures calls for an explanation. In strong inversion, the saturated drain current increases steadily whereas in weak inversion the current displays a quasi-exponential behavior. Avalanche breakdown is not the reason of course for the drain voltage is too low. The explanation is related to the impact of the threshold voltage on the pinch-off voltage. The mechanism is illustrated by means of Fig. 5.17, which takes advantage of the graphical construction introduced in Chapter 3. Left, we consider a large gate voltage so that strong inversion prevails. Right, the opposite holds true. Hatched areas represent the drain currents divided by beta as explained in Chapter 3. Since the drain voltages and gate lengths are identical in the two figures, the impact of the drain voltage on the threshold voltages is the same. Increasing V_{DS} shifts V_{To} downwards as illustrated by

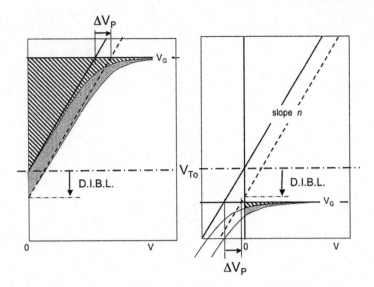

Fig. 5.17 The impact of DIBL on the drain current in strong (*left*) and weak (*right*) inversion is illustrated by means of the graphical construction presented in Chapter 3. The current increases almost linearly *left*, exponentially *right*

the two thick equal lengths arrows visible in both figures. Grey areas represent the concomitant increases of the drain currents. In strong inversion, the current grows almost linearly. In weak inversion, though the current is small, the relative increase is much larger because currents encompass a region where V_T varies exponentially.

5.5 Evaluation of g_x/I_D Ratios

The g_m/I_D and g_d/I_D ratios require to evaluate the derivatives of $\log(I_{Du})$ with respect to V_{GS} and V_{DS}. Both derivatives can be derived from the general expression:

$$\frac{g_x}{I_D} = \frac{d}{dV_x} \log(I_{Du}) = \frac{d}{dV_x} \log(i) + \frac{d}{dV_x} \log(I_{Su}) \tag{5.13}$$

where:

$$\log(I_{Su}) = \log(I_{Suo}) - \log(\theta(i)) \tag{5.14}$$

Thus:

$$\frac{g_x}{I_D} = \frac{1}{i}\frac{di}{dV_x} - \frac{1}{\theta}\frac{d\theta}{dV_x} + \frac{1}{I_{Suo}}\frac{dI_{Suo}}{dV_x} \tag{5.15}$$

which can be rewritten also as follows:

$$\frac{g_x}{I_D} = \left(1 - \frac{i}{\theta}\frac{d\theta(i)}{di}\right)\frac{1}{i}\frac{di}{dV_x} + \frac{1}{I_{Suo}}\frac{dI_{Suo}}{dV_x} \tag{5.16}$$

5.5 Evaluation of g_x/I_D Ratios

For the evaluation of the differential of the log of the normalized drain current, we take advantage of Eqs. 4.8 and 4.21–4.23 (remember V_P, V_S and V_D are defined with respect to the substrate):

$$\frac{1}{i}\frac{di}{dV_x} = \frac{1}{U_T}\left[\frac{1}{1+q_F+q_R}\frac{dV_P}{dV_x} - \frac{q_F}{i}\frac{dV_S}{dV_x} + \frac{q_R}{i}\frac{dV_D}{dV_x}\right] \quad (5.17)$$

5.5.1 The g_m/I_D Ratio

Equations 5.16 and 5.17 boil down to the expression below in the common-source configuration, since I_{Suo} doesn't depend on V_{GS}:

$$\frac{g_m}{I_D} = \frac{1}{U_T}\frac{1}{1+q_F+q_R}\left(1 - \frac{i}{\theta}\frac{d\theta(i)}{i}\right)\frac{dV_P}{dV_G} \quad (5.18)$$

which, can be rewritten as follows according to Eq. 4.14:

$$\frac{g_m}{I_D} = \frac{1}{nU_T}\frac{1}{1+q_F+q_R}\left(1 - \frac{i}{\theta}\frac{d\theta(i)}{di}\right) \quad (5.19)$$

In weak and in moderate inversion, the g_m/I_D ratio can be further simplified for mobility degradation must not be considered. This gives birth to the expression put to use by the acquisition algorithm:

$$\frac{g_m}{I_D} = \frac{1}{nU_T}\frac{1}{1+q_F+q_R} \quad (5.20)$$

In strong inversion, the evaluation of the derivative inside the parenthesis can be implemented by means of the MATLAB *polyval* and *polyder* instructions:

$$\frac{i}{\theta}\frac{d\theta(i)}{di} \Rightarrow \frac{polyval([polyder(PSu)\ 0], i)}{polyval(PSu, i)} \quad (5.21)$$

The denominator makes use of the polynomial counterpart of the 'theta' function to return θ (the polynomial *PSu* is derived from the **global variable** PolyN or PolyP -). The *polyder* instruction in the numerator takes care of the derivative of *PSu* with respect to i. The zero after the *polyder* instruction increments the order of the derivative to multiply the result by the normalized drain current i.

Figure 5.18 compares predicted to 'exact' g_m/I_D's considering the 100 nm N-channel transistor (the 'exact' data are obtained by taking the numerical derivative of the log of the 'semi-empirical' drain current). The difference between dashed and crossed lines in strong inversion illustrates the impact of mobility degradation predicted by Eq. 5.21. In weak and moderate inversion, g_m/I_D is not affected legitimating the assumptions made in the acquisition algorithm.

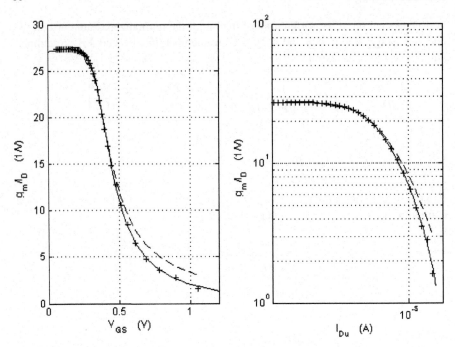

Fig. 5.18 Exact (*plain lines*) and model (*crosses*) g_m/I_D ratios versus the gate-to-source voltage V_{GS} (*left*) and unary drain current (*right*) considering the grounded source 100 nm N-channel transistor. The drain-to-source voltage is equal to 0.6 V (MATLAB fig518.m)

The three next figures illustrate the influence of the gate length, the drain-to-source voltage and back-bias on the semi-empirical and model-driven g_m/I_D ratios. In Fig. 5.19, the gate length is expanded from 0.1 to 4 μm. The smaller g_m/I_D of the 0.1 μm transistor in weak inversion reflects the larger slope factor illustrated by Fig. 5.8 that is characteristic of short channel devices. Similarly, the larger gate voltages required by the short channel device in moderate and strong inversion result from the reverse short channel effect mentioned under Fig. 5.9.

In the right figure, the impact of the gate length on the mobility degradation is clearly visible. The decay of g_m/I_D is much faster with the short channel device. Both g_m/I_D's are compared to the asymptotic construction put to use in Fig. 4.15 for the ideal transistor. The large difference with respect to the ideal transistor, even with long channel devices, is a clear warning not to infer specific currents from measurements based on the intersection of the strong and weak inversion asymptotes.

The two plots of Fig. 5.20 illustrate the influence of the drain voltage. The little impact V_{DS} has on the slope factor is corroborated by the almost similar g_m/I_D ratios in weak inversion. When the transistor is not saturated ($V_{DS} = 0.1$ V), g_m/I_D collapses very rapidly.

Figure 5.21 shows the influence of back-bias on the g_m/I_D ratio. The left side illustrates the anticipated threshold voltage and slope factor increase associated with

5.5 Evaluation of g_x/I_D Ratios

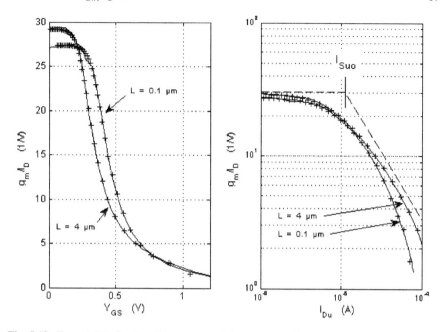

Fig. 5.19 Exact (*plain lines*) and compact model (*crosses*) g_m/I_D ratios versus gate-to-source voltage V_{GS} (*left*) and unary drain current (*right*) considering 0.1 and 4.0 μm gate lengths. The source is grounded and the drain-to-source voltage equal to 0.6 V (MATLAB fig519.m)

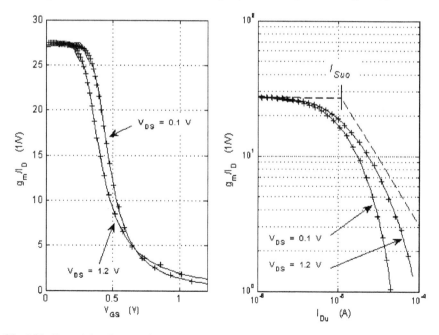

Fig. 5.20 Exact (*plain lines*) and compact model (*crosses*) g_m/I_D ratios versus the gate-to-source voltage V_{GS} (*left*) and unary drain current (*right*) considering a non-saturated ($V_{DS} = 0.1$ V) and a saturated transistor ($V_{DS} = 1.2$ V). The source is grounded and the gate length equal to 100 nm (MATLAB fig520.m)

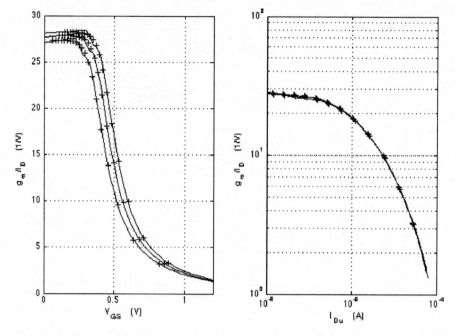

Fig. 5.21 Exact (*plain lines*) and compact model (*crosses*) g_m/I_D ratios versus the gate-to-source voltage V_{GS} (*left*) and unary drain current (*right*) considering three source voltages equal to 0, 0.4 and 0.8 V (*left to right*). The gate length is equal to 100 nm and the drain-to-source voltage 0.6 V (MATLAB fig521.m)

the growing back-bias voltage. In the right side, the curves merge practically in strong inversion (MATLAB fig521.m).

Figure 5.22 shows a magnified view of the g_m/I_D of the 100 nm N-channel transistor in weak and moderate inversion for V_{DS} equal to 0.6 V. The figure illustrates the 'filtering' effect of the compact model mentioned earlier. The model ignores the small dip near 2.7 V, which is probably due to side current.

5.5.2 The g_d/I_D Ratio

The drain conductance over drain current ratio g_d/I_D derived from Eqs. 5.16 and 5.17 boils down to the expression below where the influence of the drain voltage on the slope factor n has been neglected for it is small compared to the influence of V_{To}. When the transistor is saturated, the impact of the drain voltage is reflected by the sensitivity factor S_{VTo} and by the derivative of the log of the specific current.

$$\frac{g_d}{I_D} = \frac{1}{nU_T}\left(1 - \frac{i}{\theta}\frac{d\theta}{di}\right)\left(\frac{1}{1+q_F+q_R}|S_{VTo}| + \frac{nq_R}{i}\right) + \frac{d}{dV_D}\log(I_{Suo}) \tag{5.22}$$

5.5 Evaluation of g_x/I_D Ratios

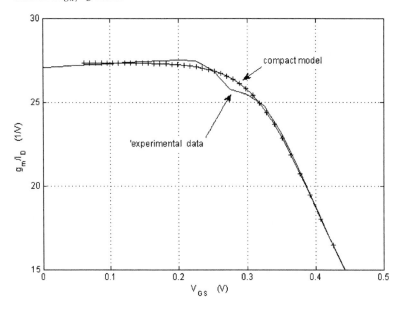

Fig. 5.22 Second order effects are ignored by the compact model (MATLAB fig522.m)

The merit of this expression is that it separates clearly the contributions of mobility degradation (the first parenthesis), D.I.B.L. (the first term in the second parenthesis), de-saturation (the second term in the second parenthesis) and C.L.M. (the last term). The impact every item has on the reciprocal of g_d/I_D, can be assessed separately thus. The point is illustrated in Figs. 5.23 and 5.24, which represent the Early voltages[5] of the N-channel transistors. In the first, V_{GS} is equal to 0.3 V (moderate inversion), in the second 0.6 V (strong inversion). Both figures report results obtained with two gate lengths: 0.1 μm left and 1 μm right. Curve (1) represents the Early voltage without D.I.B.L and C.L.M terms. As soon as the transistor enters saturation, the output conductance gets very small, almost zero, making the Early voltage very large. The transistor becomes practically an ideal current source like in the C.S.M. When second order effects are introduced, the picture changes drastically. Curve (2) shows the influence of D.I.B.L without C.L.M, whereas curve (3) combines the two. With the 1 μm transistor, the Early voltage in saturation is fixed essentially by C.L.M. The impact of D.I.B.L. is almost negligible

---

[5] The Early voltage is defined generally as the voltage where the tangent to the $I_D(V_{DS})$ characteristic crosses the horizontal axis. The Early voltage considered here is the difference between the aforementioned crossing point and the actual drain-to-source voltage. This makes I_D/V_A identical to g_d. When the Early voltage is large, the two definitions coincide more or less, but this doesn't hold true with short channel devices. In weak inversion, the zero crossing may be located even to the right of the origin owing to the exponential characteristic of the drain current like in Fig. 5.16. The Early voltage would be negative with the first definition.

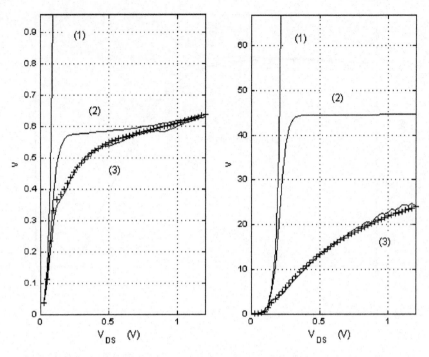

Fig. 5.23 Cumulated contributions to the Early voltage predicted by Eq. 5.23, considering a grounded 100 nm N-channel transistor (*left*) and a 1 μm (*right*). Crosses illustrate the actual semi-empirical Early voltage. The gate-to-source voltage is equal to 0.3 V (MATLAB gdID.m)

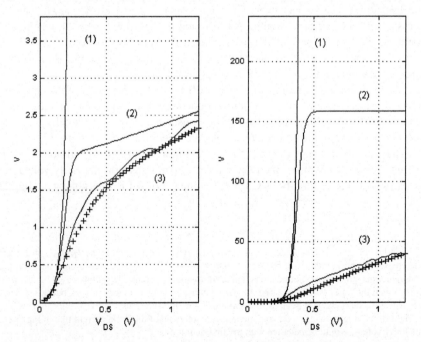

Fig. 5.24 Cumulated contributions to the Early voltage predicted by Eq. 5.23, considering a grounded 100 nm N-channel transistor (*left*) and a 1 μm (*right*). Crosses illustrate the actual semi-empirical Early voltage. The gate-to-source voltage is equal to 0.6 V (MATLAB gdID.m)

in strong inversion, small but not negligible in moderate inversion. With the 0.1 μm transistor, the opposite holds true. D.I.B.L overwhelms C.L.M.

The line consisting of crosses in the two figures illustrates the Early voltage predicted by the semi-empirical model. The 'semi-empirical' and model-driven Early voltages are more similar in moderate than in strong inversion. The difference increases with longer gate lengths owing to the increasing noise that comes with the drastic reduction of the derivatives of I_{Suo}. A more accurate approach is considered in the next chapter that does not require the derivatives of V_{T0} and I_{Suo}.

5.6 Conclusions

Drain currents predicted by the model of Chapter 4 are very similar to real drain currents. Can one extend the compact model to real transistors? The answer is yes prided some conditions are fulfilled. The model reproduces reasonably well real $I_D(V_{GS})$ characteristic even those of short-channel devices as long as the source and drain voltages are kept constant. Not only drain currents, but also g_m/I_D and g_d/I_D ratios can be reconstructed with acceptable accuracy. As soon as the drain or source voltage are modified, all parameters must be updated.

A parameter extraction algorithm is set up evaluating the slope factor, the threshold voltage, the specific current and a polynomial fit taking care of mobility degradation. The result brings about a number of interesting observations highlighting the impact of short channel effects on the parameters of the compact model, namely D.I.B.L and C.L.M.

The simplicity of the model lays down the grounds for analytical expressions. These allow performing sizing without the need to explore blindly wide ranges of drain currents. The idea is to control MOS transistors by means of variables like the normalized drain current or the forward mobile charge density q_F. A first example is considered is the next chapter concerning the sizing the real I.G.S. The method takes advantage of few parameters instead of complex advanced models with large numbers of parameters.

Chapter 6
The Real Intrinsic Gain Stage

In Chapter 1, the Intrinsic Gain Stage was sized in strong and weak inversion and, in Chapter 4, in moderate inversion. Only gradual channel models were utilized. The extension of the E.K.V model to short channel devices considered in Chapter 5 paves the way towards the sizing of real Intrinsic Gain Stages.

6.1 The Dependence on Bias Conditions of the g_m/I_D and g_d/I_D Ratios (MATLAB fig061.m)

Before undertaking the sizing, let us look to the dependence of g_m/I_D and g_d/I_D on the gate-to-source and drain-to-source voltages taking advantage of 'semi-empirical' data instead of models. The ratios are derived from the numerical derivatives with respect to the gate and the drain voltages of the log of the drain currents (see Annex A1.1 for more details regarding the derivatives).

Figures 6.1 and 6.2 display respectively constant contour plots of g_m/I_D and the reciprocal of g_d/I_D versus V_{GS} and V_{DS} considering a N-channel transistor with a gate length of 0.5 µm. In the middle of the first plot, the contours narrowing illustrates the rapid roll-off of g_m/I_D in the moderate inversion region. The drain voltage has little influence except in the upper left corner where the transistor is not saturated. In the second plot, which displays the Early voltage V_A, the commonly accepted idea that the extrapolated drain currents converge more or less to a single point on the V_{DS} axis is defeated. The only region where the Early voltage does not depend practically on the gate voltage is weak inversion.

The combination of the two plots yields the intrinsic gain for:

$$|A| = \frac{g_m}{g_d} = \frac{g_m}{I_D} \cdot \frac{I_D}{g_d} = \frac{g_m}{I_D} \cdot V_A \qquad (6.1)$$

Large g_m/I_D ratios and Early voltages are required to achieve sizeable gains. The first is synonymous of moderate and weak inversion. The second implies strong inversion and large drain-to-source voltages. Figure 6.3 shows that the best performances are obtained in moderate and weak inversion, where large g_m/I_D's

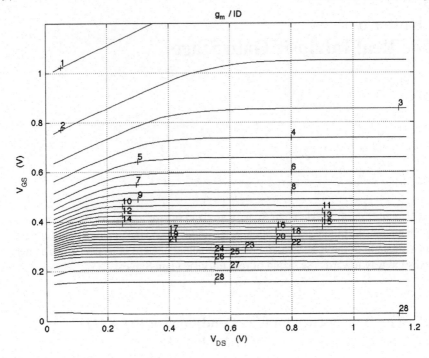

Fig. 6.1 g_m/I_D contours versus drain and gate voltages of a grounded source N-channel MOS transistor having a gate length equal to 0.5 μm

overcome poor Early voltages boosting the intrinsic gain up to 200. The gain is strongly influenced by the gate length; it ranks from 15 for 100 nm gates to more than 1,000 with 4 μm gates.

6.2 Sizing the I.G.S with 'Semi-empirical' Data

Like in previous chapters, our objective is to evaluate drain currents and sizes enabling to design Intrinsic Gain Stages that achieve a prescribed gain-bandwidth product. As stated earlier, the sizing methodology requires to have at one's disposal the $(g_m/I_D)^*$ ratio of a transistor that has the same gate length, same source and drain voltages as the transistor making out the I.G.S and a known width W^*. This ratio is obtained by taking the derivative with respect to the gate voltage of the log of the 'reference' drain current $I_D{}^*$:

$$\left(\frac{g_m}{I_D}\right)^* = \frac{1}{I_D^*}\frac{dI_D^*}{dV_G} = \frac{d}{dV_G}\log\left(I_D^*\right) \tag{6.2}$$

6.2 Sizing the I.G.S with 'Semi-empirical' Data

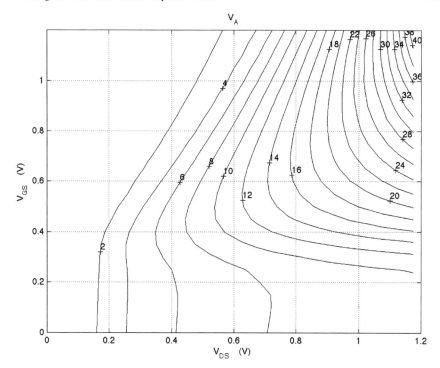

Fig. 6.2 Early voltage or I_D/g_d contours versus drain and gate voltages of a grounded source N-channel MOS transistor having a gate length equal to 0.5 μm

6.2.1 Sizing the I.G.S Loaded by a Constant Total Capacitance

In the semi-empirical method, the reference g_m over I_D is derived from measurements performed on physical transistors or reconstructed characteristics derived from advanced models. No model is put to use. The strategy is recalled in Fig. 6.4. The drain currents achieving the desired gain-bandwidth product are devised from the ratio of the transconductance g_m over the reference $(g_m/I_D)^*$, where g_m is equal to ω_T times the output capacitance C. The widths W follow from the proportionality widths – drain currents.

The MATLAB file below illustrates the method. We consider an I.G.S loaded by a 1 pF capacitor that is supposed to achieve a transition frequency of 100 MHz. The gate length, the gate-to-source voltage, the drain-to-source voltage and the source-to-substrate are listed in the first paragraph of the file. The second paragraph shares the persistent 4D arrays listed under the ***global*** instruction (for more details consult Annex 1). In the first line of the 3d paragraph, the 'reference' drain current I_{Du} is derived from the global variable IDRAINn (the index 'u' stands for unary transistors whose W/L is equal to one). Further, the reference matrix $(g_m/I_D)^*$, named gmID, is set up by taking the derivative of the log of I_{Du}. The computation is performed in two steps. We evaluate the differences between consecutive rows of the log of the

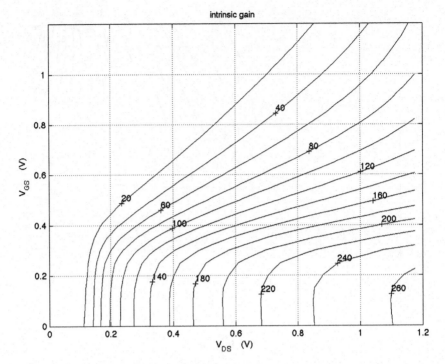

Fig. 6.3 Intrinsic gain contours versus drain and gate voltages of a grounded source N-channel MOS transistor having a gate length equal to 0.5 μm

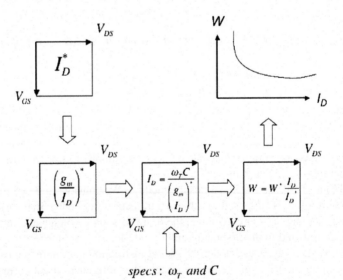

Fig. 6.4 Semi-empirical sizing method of the Intrinsic Gain Stage. The *squares* represent matrices whose rows and columns are controlled by variables associated to the *arrows*

6.2 Sizing the I.G.S with 'Semi-empirical' Data

drain current matrix by means of the ***diff*** instruction to begin with (if the matrix were transposed, the derivatives would be taken with respect to V_{DS}, yielding g_d/I_D). Then, the result $(g_m/I_D)_1$, called gmID1, is interpolated to recover the length of the original drain current vector (remind ***diff*** instructions curtail matrices by one row). Finally, the drain currents and gate widths achieving the desired gain-bandwidth product are evaluated in the fourth paragraph.

```
% 1 data
fT = 1e8;                            % Hz
C = 1e-12;                           % pF
VDS = .25: .25: 1;                   % V
VS = 0;                  % V
UG = (.1: .025: 1.2)';               % V
zG = length(UG);
L = .5;                  % µm
% 2 compute
global LL IDRAINn GMn GDSn CGGn
lg = find(LL==L);
UG = (.1: .025: 1.2)';  zG = length(UG);
vgs = round(40*UG + 1);
vds = round(40*VDS + 1);
vs = round(10*VS + 1);
% 3 construct IDu and gm/ID matrices
IDu = .1*L*squeeze(IDRAINn(lg,vgs,vds,vs));
VG = UG(:,ones(1,length(VDS)));
gmID1 = diff(log(IDu))./diff(VG);
UG1 = .5*(UG(1:zG-1) + UG(2:zG));
[X,Y] = meshgrid(VDS,UG1);
gmID = interp2(X,Y,gmID1,VDS,UG,'cubic');
% 4 size
gm = 2*pi*fT*C;
ID = gm./gmID;
W = L*ID./IDu;
% 5 gain
A = squeeze(GMn(lg,vgs,vds,vs))./squeeze...
(GDSn(lg,vgs,vds,vs));
```

Figure 6.5 displays a series of gate widths achieving the desired gain-bandwidth product considering four drain voltages V_{DS} from 0.25 to 1 V. The rapid increase of the gate widths at low drain currents denotes clearly the onset of weak inversion. No implementation is possible below a minimal current. The gate voltages V_{GS} and gains A are shown also. As stated earlier, gain is largest in weak inversion. In strong inversion, gate widths drop more or less like the reciprocal of the drain current, but when V_{GS} exceeds 0.5 V larger widths than expected are needed owing to mobility degradation.

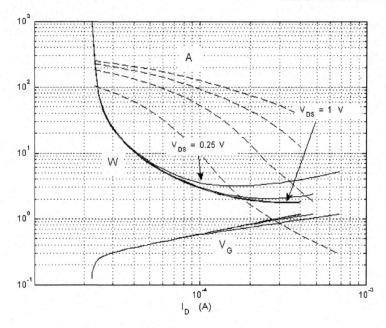

Fig. 6.5 Plot representing the gate widths W (in μm), gate-to-source voltage V_G (V) and gain A versus drain current of an N-channel I.G.S loaded by a 1 pF capacitor achieving a transition frequency of 100 MHz. The transistor has a gate length of 0.5 μm. The source is grounded and the drain-to-source voltage varies from 0.25 until 1 V in steps of 0.25 V (MATLAB fig065.m)

Figure 6.6 shows the influence of the gate length on the gate width, gate-to-source voltage and gain when L is stepped down from 0.500 to 0.160 and 0.100 nm. The influence on the gain is huge.

In the examples of Figs. 6.5 and 6.6, the frequency is supposed to be low enough to ignore the carrier's transit time in the channel. Because quasi-stationary prevails, all parameters are constants. When the transition frequency increases, there is a limit beyond which things begin to change. A landmark checking whether quasi-stationarity (q.s) conditions are likely to be met is desirable. A commonly advocated marker is the angular frequency represented by the ratio of the transistor's transconductance over its input capacitance $2\pi f_{nqs}$. When it is attained, the *current* gain of the I.G.S is equal to 1. Generally, one considers that quasi-stationarity holds as long as the frequency stays one order of magnitude below f_{nqs} (Tsividis 1999). The question is worth considering when f_T gets much larger like in Fig. 6.7, where the transition frequency has been pushed up to 1 GHz. With the 0.5 μm gate length, the f_{nqs} landmark illustrated by means of crosses lies entirely below the tenfold transition frequency limit illustrated by the thick horizontal line. To achieve the desired gain-bandwidth product, the gate length must be shortened. The 100 nm gate length fulfills quasi-stationarity even in weak inversion but the loss of the gain caused by poor Early voltages is not worth the tiny reduction of current shorter channel allows. The 160 nm transistor is a better candidate. The gain is around 30–40 while quasi-stationarity can be achieved in moderate inversion (Fig. 6.7).

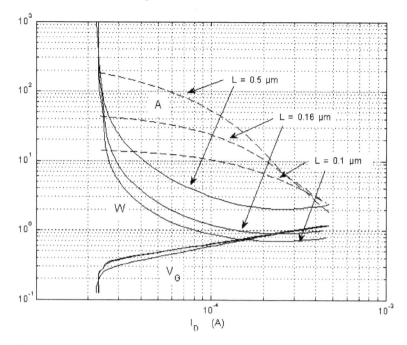

Fig. 6.6 Same experiment as in Fig. 6.5 illustrating the influence of the gate length (MATLAB fig066.m)

6.2.2 Introduction of Extrinsic Capacitances

As the gate width increases, the extrinsic junction capacitances increase too. Because the drain junction parallels the output capacitance, the total load capacitance gets larger. Keeping the capacitive load constant as we did so far, boils down to lowering the budget left over for the external load. To keep the external load unchanged, one must sum up the nominal load capacitance and the drain junction capacitance. This worsens the requirements regarding the transconductance and may lead to substantial differences, especially in low-power circuits. To evaluate the impact of drain junction parasitics, we take a closer look first to the junction capacitances.

The parasitic junction capacitances under the contact regions C_{JS} and C_{JD} illustrated in Fig. 6.8 make up a substantial part of the parasitic load of the I.G.S. The dimensions of the junctions are fixed by the technology, namely the contact holes. In the technology we consider, junctions may not be thinner than a few tenths of a micron. Though this is more than the minimum tolerated gate length, one should not forget that the capacitance of both, N and P type junctions, are still 10–15 times smaller than the gate oxide capacitance. Not only vertical, but also peripheral junction capacitances must be considered moreover. For what concerns the periphery, a distinction must be made between two regions: the region surrounding the junction outside the active region of the MOS transistor and the region facing the implanted

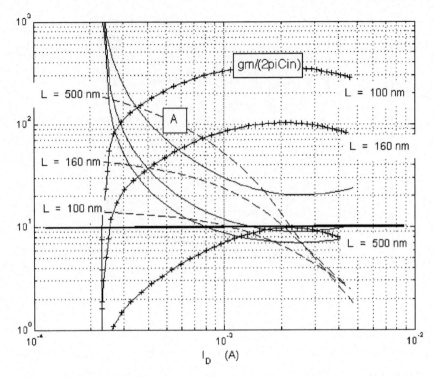

Fig. 6.7 We consider in this figure the same gate lengths as in Fig. 6.6, but fT is pushed up to 1 GHz. The curves consisting of crosses represent fmax (in GHz). Quasi-stationarity implies that the latter lie above the thick horizontal line representing ten times the I.G.S. transition frequency

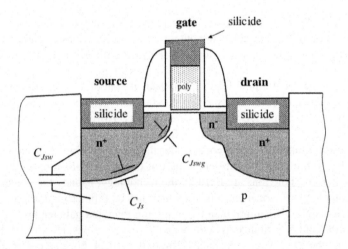

Fig. 6.8 The various contributions to the 'extrinsic' junction capacitance

6.2 Sizing the I.G.S with 'Semi-empirical' Data

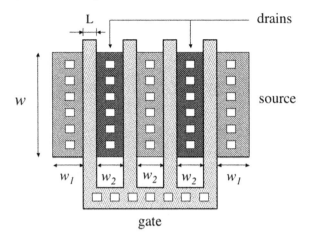

Fig. 6.9 Layout of a multigate transistor with a total gate width equal to four times w

zones that implements the transition from diffusion to channel. We call the first C_{Jsw} for side-wall peripheral capacitance, the second C_{Jswg} since it relates to the gate-side. The capacitance per unit length of the second is generally somewhat larger than that at the external periphery owing to larger amounts of impurity concentrations. Typical values for C_{Jsw} and C_{Jswg} are of the order 10^{-16} and 3×10^{-16} F/μm.

Transistor partitioning offers means to reduce the impact of the junction capacitances. The idea is to divide the transistor in smaller devices connected in parallel as shown in Fig. 6.9. Compare for instance two implementations of a same transistor, one making use of a single gate and one consisting of two halved transistors in parallel that share a common drain junction. While the capacitance of the source junction doesn't change practically, the drain junction capacitance is halved and the side-wall capacitance substantially reduced. Partitioning not only reduces the junction capacitances but also decreases the series resistance of the gates of every sub-transistor, an essential feature for high frequency applications (Grabinski et al. 2006).

The junction capacitances of partitioned transistors is given by the sum:

$$C_J = A_J C_J + P_{sw} C_{Jsw} + P_{swg} C_{Jswg} \qquad (6.3)$$

where A_J, P_{sw} and P_{swg} represent respectively the area, side-wall and gate-side lengths of the junction. Typical values for w_1 and w_2 are respectively 0.45 and 0.35 μm. For inner drain junctions, one has:

$$\begin{aligned} A_{JD} &= 0.5 \, N \cdot w \cdot w_2 \\ P_{swD} &= N \cdot w_2 \\ P_{swgD} &= N \cdot w \end{aligned} \qquad (6.4)$$

whereas, for source junctions:

$$A_{JS} = w\,(2w_1 + (0.5\,N - 1)\,w_2)$$
$$P_{swS} = 2\,(w + 2w_1) + (N - 2)\,w_2 \qquad (6.5)$$
$$P_{swgS} = N \cdot w$$

When the width W does not justify partitioning, Eqs. 6.4 and 6.5 must be replaced by the expressions below without D and S indices since nothing differentiates the source from the drain:

$$A_J = W \cdot w_1$$
$$P_{sw} = W + 2w_1 \qquad (6.6)$$
$$P_{swg} = W$$

The benefit offered by partitioning is illustrated by the curves displayed in Fig. 6.10. These compare the source and drain junction capacitances of multi-stripe implementations to the capacitance C_{J1} of a single stripe transistor having the same total width W. The horizontal axis represents the total gate width divided by the largest tolerated width $wmax$ of every sub-transistor. The ratio is equal to one until W exceeds $wmax$. Every sub-transistor has a width that is a fraction between 50% and 100% of $wmax$. Notice that the drain junction capacitance drops by almost 30% as soon as the transistor is divided in two parts. The source capacitance decreases more slowly than the drain capacitance owing to the outer junctions.

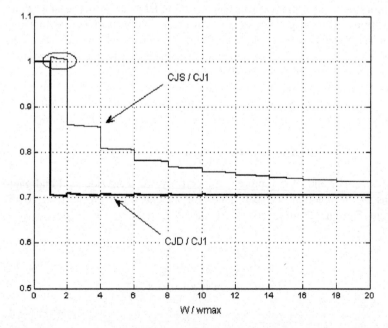

Fig. 6.10 Partitioning reduces the source and drain junction capacitances with respect to single-stripe transistors

6.2.3 Sizing the I.G.S Loaded by a Constant Load Capacitance

Sizing the Intrinsic Gain Stage while taking into account the parasitic drain junction capacitance is done by following the same procedure as above but requires to repeat the sizing procedure a few times to take care of the increasing capacitive load associated with the widening gates. The parasitic capacitance inferred from the width obtained at the end of the first run is added to the nominal output capacitance. This requires a slightly larger transconductance and gives way to new drain currents, gate widths, etc. After a few runs, convergence is reached generally. In weak inversion, a point may be reached however beyond which I_D starts to grow instead of decreasing as illustrated in Fig. 6.11. Deep in weak inversion, the width of the transistor is getting so large that the parasitic drain junction is overwhelming progressively the nominal load capacitance. It is clear that the optimum lies else, in the middle of the moderate inversion region. In the example of Fig. 6.11, a drain current of 400 µA is a good choice. The gain is not far from 40 and the non-quasi-stationarity landmark f_{nqs} still larger than ten times the transition frequency.

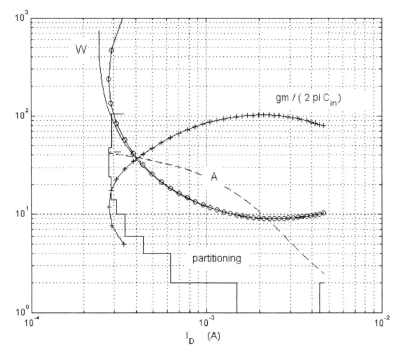

Fig. 6.11 Impact of the parasitic drain junction capacitance on the 160 nm I.G.S. considered in Fig. 6.7. The partitioning factor N illustrated by the broken line is fixed by the 10 µm maximum width imposed to every sub-transistor (MATLAB fig611.m)

6.3 Model Driven Sizing of the I.G.S.

We now undertake sizing considering the compact model instead of 'semi-empirical' data. The presentation is divided into two parts: first, the evaluation of W and I_D, second, the Intrinsic Gain A.

6.3.1 Sizing W and ID (MATLAB fig612.m)

The model-driven method makes use of the same equations as the semi-empirical but implements the reference $(g_m/I_D)^*$ differently. In the semi-empirical approach, the ratio was evaluated numerically. In the model-driven, it is derived from the parameters n, V_{To}, I_{Suo} and the *Theta* function. One takes advantage moreover of the fact that the model-driven method offers the possibility to focus sizing on a well-defined region or mode of operation. As a result, one can perform sizing while trading gain against low power consumption by selecting appropriate g_m/I_D's. Other pointers than the transconductance over drain current ratio can be put to use as well. The normalized drain current and the forward normalized mobile charge density q_F are attractive contenders for they measure how deep transistors operate in moderate, weak or strong inversion. For instance, q_F equal to one lies middle in the moderate inversion region, q_F's smaller than 0.1 correspond to weak inversion and q_F's larger than 10 to strong inversion. The fact that q_F is not a voltage or a current doesn't matter; once sizing is completed the variable disappears like in the parametric method illustrated by Eq. 4.24.

The excerpts from the MATLAB file below show an example. First, the compact model parameters are extracted from *global* variables having the same names[1]. We define a q_F logspace vector that encompasses the moderate inversion region. This leads to the evaluation of the pinch-off voltage V_P (remind V_S is equal to 0), paving the road towards the gate-to-source voltages V_{GS} and the normalized reverse mobile charge density vector q_R. The normalized drain current i follows. The sizing algorithm is put to use in three steps. To begin with, we evaluate the unary drain current and g_m/I_D ratio without considering mobility degradation nor parasitic drain junction capacitance.

```
global LL nN VToN ISuoN PolyN
...
n = nN(vds,vs,lg);
VTo = VToN(vds,vs,lg);
ISuo = ISuoN(vds,vs,lg);
P = squeeze(PolyN(vds,vs,lg,:));
...
```

---

[1] These consist of arrays controlled by the drain-to-source voltage V_{DS}, the source-to-substrate voltage V_S and the gate length L (see Annex 1).

6.3 Model Driven Sizing of the I.G.S.

```
qF = logspace(-1.8,.8,30);
VP = UT*(2*(qF-1) + log(qF));
VGS = n*VP + VTo;
qR = invq((VP-VDS)/UT);
i = qF.^2 + qF - qR.^2 - qR;
IDu = ISuo.*i;
gmID1 = 1./(n*UT.*(1+qF+qR));
```

The drain current I_{D1} and width W_1 vectors achieving the desired gain-bandwidth product are obtained then like in Section 6.2.1. The result is illustrated by the dashed curve of Fig. 6.12 connecting the weak and strong inversion asymptotes represented by the two thick straight lines like in Chapter 1, the vertical for weak inversion, the other for strong inversion.

During the second sizing step, we introduce mobility degradation by adding the lines below to the file. The first line evaluates I_{Su}, the second I_{Du} and the two last the g_m/I_D ratio according to Eqs. 5.19 and 5.20.

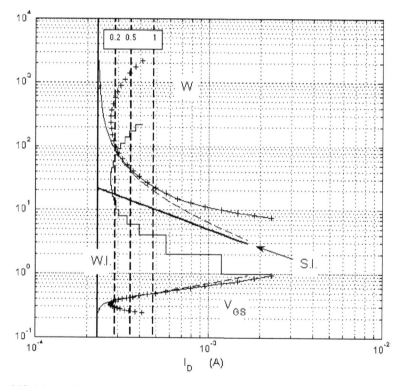

Fig. 6.12 Model-driven sizing of the 1 GHz gain-bandwidth I.G.S. considered in Fig. 6.11. The source is grounded, the drain voltage equal to 0.6 V, the gate length equal to 160 nm and the max width of partitioned transistors 10 μm. The broken line relates to the transistor partitioning. The vertical lines correspond to three qF's (MATLAB fig612.m)

```
P = PolyN(vds,:,vs,lg); ISu = ISuo*polyval(P,i);
IDu = i.*ISu;
Z = (1 - polyval([polyder(P) 0],i)./polyval(P,i);
gmIDD = gmID1.*Z;
```

The result is illustrated by the plain line curve of Fig. 6.12. In weak inversion, widths merge with those computed earlier. Mobility degradation comes to the fore only in strong inversion as the plain line curve moves away increasingly from the dashed curve.

In the third step, we introduce the drain junction parasitic capacitance paralleling the output load. Since the parasitic capacitance varies like the transistor width and the latter is a function the transconductance, the algorithm makes use of a loop. Crosses and the staircase curve illustrate the widths and partitioning factor.

```
C = Co;
for k = 1:10,
  Gm = 2*pi*fT*C;
  ID2 = Gm./gmIDD;
  WsL2 = ID1./IDu;
  W2 = WsL2*L;
  JCap2 = jctCap(L,W2,maxW,VDS); CJD2 = JCap2(:,:,1);
  N = JCap2(:,:,3);
  C = Co + CJD2;
end
```

It is clear that widths larger than 100 μm don't make sense. The currents and widths within the region delineated by the vertical dashed lines defined by q_F's respectively equal to 0.2, 0.5 represent good compromises. Naturally, moderate inversion offers the best compromise.

Notice that all the curves coincide more or less in the moderate inversion region. In this region, the widths and currents are not strongly influenced by mobility degradation nor drain parasitic capacitance. Three parameters, n, V_{To} and I_{Suo}, are enough in order to size the I.G.S in this region thus.

6.3.2 Evaluation of the Intrinsic Gain (MATLAB fig613.m)

To evaluate the intrinsic gain we multiply the transconductance over drain current ratio by the Early voltage:

$$A = \left(\frac{g_m}{I_D}\right)^* \left(\frac{I_D}{g_d}\right)^* = \left(\frac{g_m}{I_D}\right)^* V_A^* \tag{6.7}$$

taking for the reciprocal of the Early voltage the expression below:

$$\frac{g_d}{I_D} = \frac{d}{dV_{DS}} \log(I_{Du}) = \frac{d}{dV_{DS}} \log(i) + \frac{d}{dV_{DS}} \log(I_{Su}) \tag{6.8}$$

Since V_S and V_{GS} are constants, the derivative of the log of the normalized drain current boils down to the expression below derived from Eq. 5.17:

$$\frac{d \log(i)}{dV_{VS}} = \frac{q_R}{i\,U_T} - \frac{1}{nU_T}\frac{1}{1+q_F+q_R} S_{VTo} \quad (6.9)$$

The first term after the equal sign takes care of de-saturation and the second of D.I.B.L. The first vanishes as soon as the drain voltage exceeds 100 m. The remainder consists of two factors: one is the g_m/I_D ratio evaluated earlier, the other the threshold voltage sensitivity factor. Since the threshold voltage varies quasi-linearly with V_{DS}, a first order expansion of the threshold voltage suffices. The lines hereafter illustrate the evaluation of the sensitivity factor:

```
U = (0:.025: 1.2)'; zu = length(U);
P1 = polyfit(U(2:zu),VToN(2:zu,vs,lg),1);
SVTo = P1(1);
```

The second term of Eq. 6.8, which relates to channel length modulation (C.L.M.), requires computing the derivative of the log of the specific current with respect to the drain current instead of the gate voltage. To perform the derivation in the orthogonal direction, we compute the specific current considering the nominal and two adjacent drain voltages and take the averaged *diff* of the log of the specific currents.

```
Y = ISuN(vds+(-1:1),:, vs,lg);
CLM = mean(diff(log(Y)))/.025;
```

The reciprocal of the Early voltage can now be evaluated by summing the contributions of D.I.B.L and C.L.M in de-saturation:

```
gdID = qR/(UT*i) - SVTo*gmID1 + CML;
```

We can evaluate the intrinsic gain 'A' by combining the g_m/I_D found earlier with the g_d/I_D above according to Eq. 6.7. Figure 6.13 compares the model-driven gain to the 'semi-empirical' gain evaluated in Section 6.2 (the figure compares also the widths and gate voltages). Physical and model-driven approaches are equivalent with the exception of the gain in strong inversion when the transistor de-saturates. Contrarily to the 'physical' approach, the model-driven appends some features. It offers the possibility to sense the respective contribution of D.I.B.L and C.L.M. The fact that omitting the C.L.M term in the expression above doesn't change practically the gain of the 100 nm transistor is a clear confirmation that D.I.B.L is overwhelming C.L.M in short channel devices. The opposite holds true with the 4 μm transistor.

6.3.3 An Alternative Method to Evaluate the Gain (MATLAB fig615.m)

The derivatives required in order to evaluate gain introduce computation noise, in particular the derivative of the specific current. While short channel devices are not too much affected by noise, long channel encounter problems due to the smallness

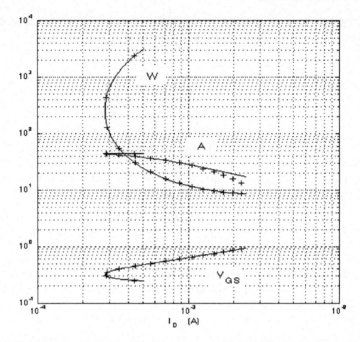

Fig. 6.13 The figure compares the width, gate-to-source voltage and intrinsic gain predicted by the model (*continuous lines*) to its semi-empirical counterparts (*crosses*). The gain-bandwidth product is equal to 1 GHz and the output capacitor equal to 1 pF like in the previous figure (MATLAB fig613.m)

of the C.L.M and D.I.B.L. contributions. Better results can be obtained when the derivatives are evaluated at a later stage. In the approach hereafter, the gain is inferred from the slope of the I.G.S transfer characteristic.

Consider once more the 160 nm grounded source transistor targeting a gain-bandwidth product of 1 GHz. We select a quiescent q_{Fo} of 0.5 that corresponds to the middle vertical dashed line represented in Fig. 6.12. The I.G.S operates clearly in moderate inversion. To find the gain we construct the transfer characteristic sweeping the drain voltage throughout the entire output range and search the correspondent gate voltages. Though the drain current does not vary for the I.G.S. is fed by a current source, the normalized drain current does for the specific current depends on the drain voltage. Dividing the constant drain current by the variable specific current paves the way to the normalized drain current, which in turn leads to the normalized mobile charge density q_F and pinch-off voltage vectors. The gate voltage follows since the slope factor and threshold voltage dependence on the drain voltage are known.

The procedure is illustrated by means of the MATLAB file hereafter, which takes into account mobility degradation and transistor de-saturation. Since we don't know the reverse mobile charge density q_R nor the degree of mobility degradation when we start, the calculation proceeds by reiterating the evaluation a few times. For the first run q_R is supposed to be equal to zero and the *theta* function equal to one.

6.3 Model Driven Sizing of the I.G.S.

After every cycle, a better approximation of the pinch-off voltage is obtained, for we reevaluate q_R by subtracting V_{DS} from V_{PS}. After a few runs, the algorithm converges. An interpolation step expressing the drain voltage as a function of the gate voltage is performed.

```
qR = 0;
TH = 1;
[X1,Y1] = meshgrid(U);
er = 1;
while er > 1e-3,
  QR = qR;
  i = IDo.*TH./(W/L*ISuo);
  qF = .5*(sqrt(1 + 4*(i+QR.^2+QR)) - 1);
  VPS = UT*(2*(qF - 1) + log(qF));
  qR = invq((VPS-VDS)/UT);
  VGS = n.*VPS + VTo;
  TH = diag(interp2(X1,Y1,ThN(:,:,vs,lg),VGS,VDS',
  'cubic'));
  er = max(abs(1 - QR./qR));
end
```

The transfer characteristic is illustrated in Fig. 6.14 together with the forward and reverse mobile charge densities q_F and q_R. When V_{DS} is equal to 0.6 V, q_F is equal

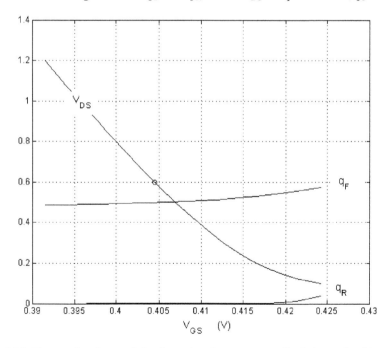

Fig. 6.14 The transfer characteristic, forward and reverse normalized charge densities of the N-channel transistor achieving a gain-bandwidth product of 1 GHz. The supply voltage VDD is equal to 1.2 V. The steady state point corresponds to the *circle*

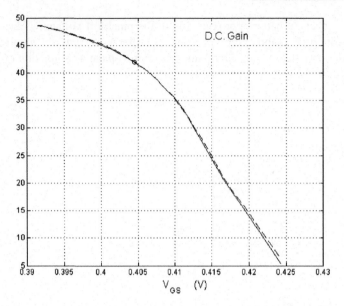

Fig. 6.15 D.C. gain of the I.G.S. considered in the previous figure. The characteristic predicted by the model is represented by means of *plain lines* and the semi-empirical by *dashed lines*

to 0.5. The transistor is saturated and q_R negligible. As the drain voltage lessens, the transistor progressively de-saturates, and q_R begins to increase pushing q_F upwards to keep the drain current constant.

The gain of the I.G.S can be derived from the slope of the transfer characteristic. The result shown in Fig. 6.15, compares the gain predicted by the model to the gain of the 'semi-empirical' model.

Notice that the current source feeding the I.G.S must not be an ideal current source necessarily. If a P- channel transistor is put in the place of the current source, the drain current becomes a function of the output voltage. Once the drain current and voltage known, the construction of the transfer function proceeds like above.[2]

6.3.4 A Simplified Sizing Procedure

It is clear that moderate inversion offers the most interesting compromise as far as power consumption and transistor widths. In moderate inversion, the impact of mobility degradation is small and may be neglected generally. Sizes and drain currents can be evaluated in a straightforward manner as long as the transistor is saturated. The process is illustrated by the flow chart of Fig. 6.16. The starting point is the

---

[2] Computing the transfer function allows to evaluate harmonic distortion.

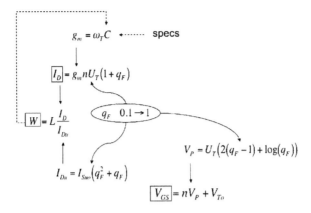

Fig. 6.16 I.G.S sizing flow chart in moderate inversion

choice of an appropriate normalized forward mobile charge density q_F scalar or vector. The drain current I_D achieving the desired gain-bandwidth product is obtained by multiplying the transconductance g_m by the reciprocal of g_m/I_D. The width W is found by dividing I_D by the unary drain current, which is equal to I_{Suo} times the normalized drain current. Eventually, the parasitic drain capacitance is added to C, which implies to reiterate the evaluation of I_D and W. The gate voltage is derived from the pinch-off voltage V_P.

6.4 Slew-Rate Considerations

Output voltage changes require to charge and discharge the capacitor loading the output terminal of the I.G.S. When the output voltage increases, the current delivered by the current source in the drain is split in two parts. A fraction charges the output capacitor while the rest feeds the transistor. When the rate at which the output voltage increases gets too large, the current feeding the transistor may dry out. The output voltage increases still but the slope dV_{out}/dt cannot exceed the limit set by the 'slewing rate' I_D/C where I_D is the DC current delivered by the current source.

The slewing rate, the gain-bandwidth product, and the g_m/I_D ratio are related, for:

$$\text{slewing rate} = \frac{I_D}{C} = \frac{g_m/C}{g_m/I_D} = \frac{\omega_T}{g_m/I_D} \quad (6.10)$$

So far, our only concern has been to lower power consumption and to get more gain. We haven't considered slew-rate. The latter however impacts the I.G.S performances since the largest slope sine waves the I.G.S. can display depends on both, the magnitude (V) and the angular frequency (ω):

$$\left(\frac{dV_{out}}{dt}\right)_{max} = \omega V \quad (6.11)$$

When combined, Eqs. 6.10 and 6.11 lead to the expression below, which must be satisfied to avoid non-linear distortion:

$$\omega V < \frac{\omega_T}{\frac{g_m}{I_D}} \qquad (6.12)$$

Sine wave peaks cannot trespass the reciprocal of g_m/I_D thus at the transition frequency. This is a severe limitation not to overlook. An I.G.S. intended to operate in a unity gain loop may require for instance enhancing the transition frequency by a large factor depending on the targeted g_m/I_D and the required dynamic range. The sizing algorithm doesn't change but the transition frequency needs to be enhanced. With low-voltage circuits fortunately, the impact is less acute for the dynamic range is necessarily small.

6.5 Conclusions

The Intrinsic Gain Stage sizing procedure described in Chapters 1 and 4 is revisited considering real transistors. 'Semi-empirical' data are considered first. The compact model introduced in Chapter 5 follows. The two yield close results.

One of the assets of the model-driven methodology is that sizing can be done in well-defined regions. The normalized forward mobile charge density offers an effective means to restrain sizing to moderate inversion whereas the semi-empirical method proceeds blindly. The model allows moreover tracing the relative contributions of second order effects, like D.I.B.L and C.L.M. Although not a major asset for sizing, the physical insight the model provides is worth mentioning.

Chapter 7
The Common-Gate Configuration

7.1 Drain Current Versus Source-to-Substrate Voltage (Matlab fig071.m)

In the common gate configuration, the gate-to-source and the drain-to-source voltages, V_{GS} and V_{DS}, vary with the source-to-substrate voltage V_S. As a result, the compact model parameters require continuing updating.

Figure 7.1 displays the drain current versus the source voltage V_S of the 100 nm N-channel transistor considered in the previous chapter taking advantage of updated n, V_{To} and I_{Suo} parameters. The gate- and drain-to-substrate voltages are constant and respectively equal to 0.9 and 1.0 V. The currents predicted by the compact model with and without mobility degradation are represented respectively by the continuous and dashed curves. Crosses represent the 'semi-empirical' drain current. When V_S is small, the impact of mobility degradation is considerable for the gate-to-source and drain-to-source voltages are large. As V_S increases, the two curves concur progressively until they merge in weak inversion giving birth to the distinctive weak inversion straight line.

The transconductance over drain current ratios of the model and 'semi-empirical' data are represented in Fig. 7.2. The model g_{ms}/I_D ratio is derived from Eq. 5.16:

$$\frac{g_{ms}}{I_D} = \left(1 - \frac{i}{\theta}\frac{d\theta}{di}\right)\frac{1}{i}\frac{di}{dV_s} + \frac{1}{I_{Suo}}\frac{dI_{Suo}}{dV_s} \tag{7.1}$$

Since the drain-to-substrate voltage V_D is constant, the derivative of the normalized drain current with respect to V_S given by Eq. 5.17, boils down to the expression:

$$\frac{1}{i}\frac{di}{dV_S} = \frac{1}{U_T}\left[\frac{1}{1+q_F+q_R}\left(\frac{dV_P}{dV_S}\right) - \frac{q_F}{i}\right] \tag{7.2}$$

which can be further simplified when the transistor is saturated:

$$\frac{1}{i}\frac{di}{dV_S} = \frac{1}{U_T}\frac{1}{1+q_F}\left(1 - \frac{dV_P}{dV_S}\right) \tag{7.3}$$

P.G.A. Jespers, *The g_m/I_D Methodology, A Sizing Tool for Low-voltage Analog CMOS Circuits*, Analog Circuits and Signal Processing, DOI 10.1007/978-0-387-47101-3_7, © Springer Science+Business Media, LLC 2010

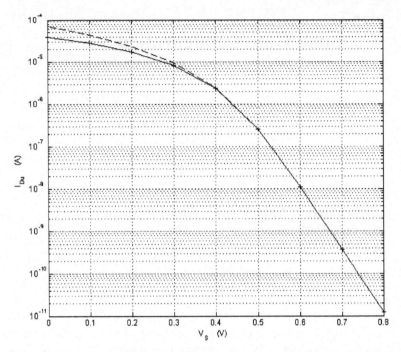

Fig. 7.1 Common gate drain current versus the source-to-substrate voltage of the unary 100 nm N-channel transistor. The drain-to-substrate voltage V_D and the gate-to-substrate voltage V_G are constant and respectively equal to 1 and 0.9 V. Crosses represent the 'exact' current obtained by mean of the semi-empirical method. The dashed and continuous lines relate to the compact model without and with mobility degradation (MATLAB: fig071.m)

In weak inversion, the factor between parentheses in Eq. 7.1 can be omitted turning the g_{ms}/I_D ratio into the expression:

$$\frac{g_{ms}}{I_D} = \frac{1}{U_T}\left(1 - \frac{dV_P}{dV_S}\right) + \frac{d\log(I_{Suo})}{dV_S} \quad (7.4)$$

If all parameters were constant, the maximum of g_{ms}/I_D would be equal to $1/U_T$ for the derivatives of the pinch-off voltage and the specific current vanish. With real transistors, this isn't the case. The pinch-off voltage varies with V_{GS} and V_{DS} for it depends on V_{To} and n. The same holds true *for* the specific current I_{Suo}. The result is a maximum g_{ms}/I_D below the theoretical limit of 38 V^{-1}. In the example, the maximum is equal to 34 V^{-1}, which corresponds to a slope factor of 1.13. Contrarily to the Charge Sheet Model, the slope factor in weak inversion is not equal to one but slightly larger owing to the influence of the drain.

7.2 The Cascoded Intrinsic Gain Stage

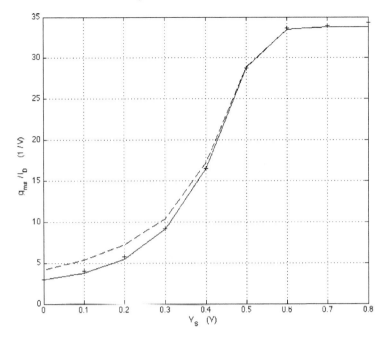

Fig. 7.2 g_{ms}/I_D ratio of the transistor of Fig. 7.1. The model is illustrated by means of the dashed and continuous lines like in Fig 7.1, the semi-empirical ratio by means of crosses (MATLAB: fig071.m)

7.2 The Cascoded Intrinsic Gain Stage

Common-gate stages are currently put to use in order to perform impedance transformations. The impedance looking into the source of the common gate transistor is a replica of the output impedance *divided* by the gain while the impedance seen at the drain is a replica of the load in the source *multiplied* by the gain. The foremost example of a circuit taking advantage of this is the cascode circuit shown in Fig. 7.3. It consists of two transistors, a grounded source and a common gate transistor. The transconductance is set by the common source stage for the same current is flowing through the two transistors. The output impedance is a replica of the output impedance of the common-source transistor times the gain A_2 of the common-gate transistor for the source of Q_2, which is also the drain of Q_1, replicates the output of the cascode divided by the gain A_2.

7.2.1 Sizing the Cascode (Matlab fig074.m)

The circuit represented in Fig. 7.3 is actually a cascoded version of the Intrinsic Gain Stage. Sizing follows similar lines. Consider a low-power low-voltage cascode

Fig. 7.3 The basic cascode configuration

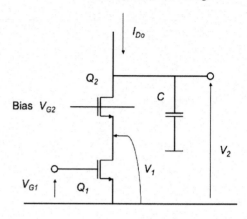

stage loaded by a 1 pF capacitor supposed to achieve a gain-bandwidth product of 100 MHz. The two transistors are saturated and their drain-to-substrate voltages V_1 and V_2 are respectively equal to 0.3 and 0.6 V. Since the source and drain voltages of both transistors are fixed, all parameters are at hand.

We confine the mode of operation of the two transistors to moderate inversion for this is the best compromise as far as gain and power consumption. We assume that the increased sensitivity of transistors operating in this mode can be counteracted by proper bias circuitry. Let us choose a gate length of 0.5 μm for both transistors and start with the sizing of Q_1. Consider a q_{F1} vector in moderate inversion, for instance from 0.1 to 2. Mobility degradation is not be taken into consideration to evaluate the unary drain current I_{Du1}. The transconductance g_{m1} is obtained by multiplying the desired angular transition frequency by the output capacitance. The drain current vector I_{D1} follows from the ratio g_{m1} over $(g_m/I_D)_1$ while the aspect ratio W_1/L_1 is obtained by dividing I_{D1} by the unary drain current I_{Du1} as usual. Consider now Q_2. Generally, one chooses for Q_2 the same width as for Q_1. Since the drain current and width of the common-gate stage are known, we evaluate the normalized drain current by dividing I_{Du2} (which is equal to I_{Du1}) by I_{Suo2}. This leads to the normalized mobile charge density q_{F2}. We can now calculate the pinch-off voltages of Q_1 and Q_2 and find the gate-to-source voltages of both transistors as well as their gate-to-substrate voltages. The procedure is repeated a few times to take care of the parasitic junction capacitance paralleling the 1 pF output load. Mobility degradation can be introduced eventually at this stage if needed.

The result shown in the left part of Fig. 7.4 displays the transistor's widths and gate-to-substrate voltages achieving the desired 100 MHz gain-bandwidth product considering normalized mobile charge densities comprised between 0.03 and 2. The lower limit of q_{F1} is clearly unpractical. The upper limit isn't interesting either for less power consumption can be attained with reasonable widths. The 20 μm width marked by a circle seems to be a good compromise. The normalized mobile charge densities q_{F1} and q_{F2} are respectively equal to 0.32 and 0.33 (clearly in moderate inversion), the drain current I_D is equal to 31.5 μA, and the gate-to-substrate voltages V_{G1} and V_{G2} are respectively equal to 0.320 and 0.671 V.

7.2 The Cascoded Intrinsic Gain Stage

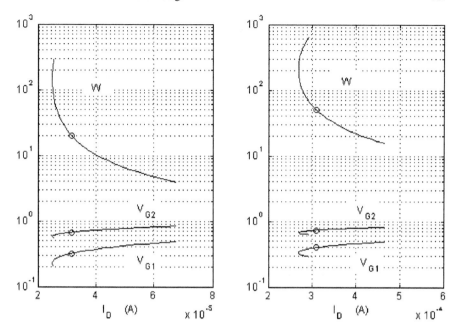

Fig. 7.4 Width and gate voltages of two cascoded Intrinsic Gain Stages loaded by a 1 pF capacitance. Left, the gain-bandwidth product is equal 100 MHz, right 1 GHz. The gate lengths are respectively 500 and 100 nm (MATLAB fig074.m)

In the right part of the same figure, the transition frequency is increased from 100 MHz to 1 GHz while the gate length is reduced from 500 to 100 nm. Widths and drain currents increase of course. For the 50 μm width marked by the circle, the drain current is equal to 310 μA, V_{G1} and V_{G2} are equal to 0.400 and 0.736 V and q_{F1} and q_{F2} nearly the same as in the left plot.

To summarize, the sizing methodology of the cascaded I.G.S. proceeds as follows: (1) fix a range of normalized mobile charge densities offering a good compromise power consumption versus sizes. (2) Evaluate the normalized drain currents and g_m/I_D's. (3) Make use of the g_m/I_D methodology to get drain currents and aspect ratios fulfilling the gain-bandwidth requirements. (4) Choose the widths of Q_1 and Q_2 that achieve low drain current and evaluate the gate-to-source voltages of the two transistors. (5) Retrieve the file to take care of the parasitic output junction capacitance paralleling the output load and mobility degradation.

7.2.2 Gain Evaluation of the Cascode (MATLAB fig075.m)

How to evaluate gain? Rather than assessing derivatives, we opt for the same approach as in Section 6.3.3, where the gain was derived from the transfer characteristic. Suppose the current I_{Do} feeding the cascode is delivered by an ideal current source like in the I.G.S. The method proceeds as follows: search input voltages that

Table 7.1 Gains of the two cascode circuits of Fig. 7.4

Type	DV_2 (mV)	DV_1 (μV)	A2 (dB)	DV_{GS1} (μV)	A1 (dB)	A (dB)
100 MHz 500 nm	2	17.8	41	0.178	40	81
1 GHz 100 nm	2	154	22.3	12.9	21.5	43.8

keep the drain current unchanged when the output voltage is slightly modified. The evaluation proceeds in two steps. First, we evaluate the voltage excursion DV_1 at the source of the common gate transistor that results from the output voltage variation DV_2. Second, we evaluate the concomitant input voltage variation DV_{GS1} of the common source transistor. The method takes advantage of interpolation techniques like in Section 6.3.4. These track accurately small signals. Suppose for instance that the drain voltage V_2 of Q_2 is incremented by 1 mV. To find the corresponding source voltage change DV_1, we set up a source test-vector V_{S1} and make use of interpolated parameters to evaluate the concomitant drain current vector I_D. The source voltage we are looking for is extracted from V_{S1} by means of a second interpolation instruction searching the source voltage that makes the drain current equal to I_{Do}. The voltage step DV_1 caused by DV_2 lies now for the hand. Knowing DV_1, we derive DV_{GS1} along similar lines. Not only we get the gain of the cascode by dividing DV_2 by DV_{GS1}, but also the gains A_1 and A_2 of the two stages making out the cascode. The gains of the two circuits are reported under Table 7.1. The fact that g_{ms} is larger than g_m explains why the gain of Q_2 is always slightly large than that of Q_1 notwithstanding back-bias.

7.2.3 The Poles of the Cascode Circuit (MATLAB fig075.m)

The sizing procedure above does not consider whether the cascode is a stable circuit or not. We acted as if the output node represents the only pole of the circuit. There is a second pole however that is related to the phase lag caused by the parasitic capacitance at the common node of Q_1 and Q_2. The capacitance at this node consists not only of the junction capacitances of Q_1 and Q_2, but also of the intrinsic source capacitance of Q_2. While junction parasitic capacitances can be evaluated easily, the intrinsic source capacitance of Q_2 is more difficult to apprehend. It is the sum of the source-to-gate and source-to-substrate capacitances plus source-to-drain intrinsic capacitance, which is negative. In practice, the total intrinsic capacitance is close to the halved gate capacitance in strong inversion. Since the non-dominant pole must lie beyond the transition frequency to yield stability, a rough estimate of the intrinsic capacitance suffices.

The frequency responses of the cascoded circuits are shown in Figs. 7.5 and 7.6. The small signal parameters of the common source and common gate stages put to use for the evaluation of the frequency responses are derived from the compact model with the exception of the intrinsic source capacitances, which is extracted from the 'exact' semi-empirical *global* variable CSSn. The phase margins justify the assimilation of both circuits to dominant pole first order systems.

7.2 The Cascoded Intrinsic Gain Stage

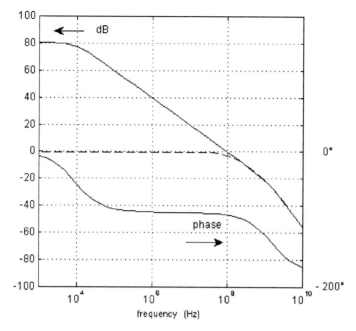

Fig. 7.5 Frequency response of the left-sided cascode circuit of Fig. 7.4. *Plain lines* represent the open-loop magnitude and phase characteristics. The *dashed lines* relates to the unity-gain configuration

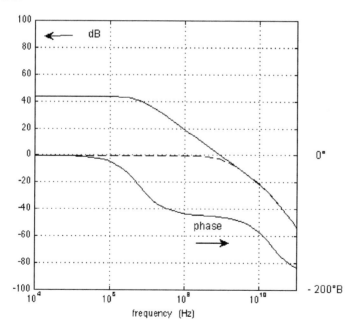

Fig. 7.6 Frequency response of the right-sided cascode circuit of Fig. 7.4. Plain lines represent the open-loop magnitude and phase characteristics. The dashed lines relates to the unity-gain configuration

Chapter 8
Sizing the Miller Op. Amp.

8.1 Introductory Considerations

Fixing currents and transistors widths of Op. Amps is a multifaceted task owing to the growing number of choices that can be made. Sizing implies hierarchy. Some objectives ought to be satisfied whichever choices. They shape the *specifications* list. A typical example is the I.G.S gain bandwidth product. Other objectives are desirable but not mandatory. They determine *attributes* like power consumption versus area. *Specifications* determine the dimensions of the g_m/I_D sizing space while *attributes* delineate optimization areas within the sizing space. The *specifications* of the Miller Op. Amp considered in this chapter are twofold: a prescribed gain-bandwidth product and an assessment regarding stability. The sizing space conforms to a two-dimensional space. Every point represents a distinct Miller Op. Amp that fulfills the same *specifications*. Low-power consumption demarcates a region within the 2D sizing space. Area minimization relates to another region. Eventually regions intersect easing choices. Whichever combination, *specifications* must be met anyway.

The axes of the sizing space play the same role as the gate voltage, drain current or normalized drain current in the I.G.S. They represent variables controlling the modes of operation of transistors or ensembles of transistors. In the Miller Op. Amp, we are going to focus on the two stages and control their behavior by means of two distinct vectors. Each vector is supposed to control transistors that have a strong impact on the fulfillment of the specifications.

8.2 The Miller Op. Amp.

The Miller Op. Amp that we consider in this chapter consists of two cascaded stages, a differential amplifier followed by a common source stage. In the circuit shown in Fig. 8.1, the first stage consists of a P-channel differential stage, the second of a N-channel common-source stage. The second stage is a true Intrinsic Gain Stage. The AC current generated by the differential pair is fed to the input of the second stage through a current mirror. Complementary transistors ease the transfer from the

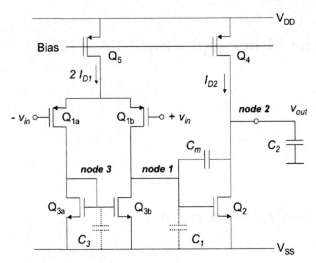

Fig. 8.1 The basic Miller Op. Amp

first to the second stage by lifting up the small signals DC level to the ground level. The choice of PMOS transistors for the first stage is credited to the better 1/f noise performances of P-channel with respect to N-MOS transistors. While the input of the Op. Amp is symmetrical, the output is asymmetrical.

The Op. Amp has two high impedance nodes marked 1 and 2, which determine two poles. These lie generally below the transition frequency and turn the Op. Amp into a second order system. Without proper action, instability is unavoidable for the amplifier is not short of additional phase lag sources. The purpose of the capacitor C_m is to change the Op. Amp into a first order system by shifting the pole associated with node 1 to low frequencies and the pole associated to node 2 beyond the transition frequency. The name of 'Miller capacitance' given to C_m is a tribute to J.M. Miller (Miller 1920) who first recognized the role of the capacitance bridging in and output terminals of inverting amplifiers.

8.2.1 Analysis of the Miller Operational Amplifier

Before sizing, a preliminary analysis of the Miller Op. Amp is carried out in order to identify the principal mechanisms controlling its frequency response. We therefore replace the amplifier by the equivalent circuit shown in Fig. 8.2, which consists of two cascaded Intrinsic Gain Stages. The first stands for the differential amplifier plus the current mirror. Since each transistor Q_1 'sees' half of the symmetrical input signal, the contributions of Q_{1a} and Q_{1b} to the overall transconductance are halved. The global transconductance g_{m1} of the first stage however is the same as that of Q_{1a} or Q_{1b} for the current mirror recombines the output currents of the differential

8.2 The Miller Op. Amp.

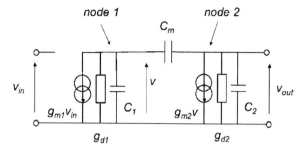

Fig. 8.2 Simplified equivalent circuit of the Miller Op. Amp

Table 8.1 Numerical values of the equivalent circuit of Fig. 8.2. The parameters g_{m3} and C_3 of the current mirror that load the first stage are not considered in the equivalent circuit. They are introduced later

g_{m1}	1.76e-04 S
g_{m2}	1.76e-03 S
g_{m3}	1.62e-04 S
g_{d1}	1.56e-06 S
g_{d2}	2.00e-05 S
C_m	5.62e-13 F
C_1	1.26c-13 F
C_2	1.04e-12 F
C_3	8.93e-14 F

stage. The output conductance g_{d1} of the first stage is the sum of the output conductances of transistor Q_{3b} and the halved output conductance of Q_{1b} for the source of the latter is connected to the source of Q_{1a}. The transconductance of the second stage is called g_{m2} and g_{d2} globalizes the output conductances of Q_2 and Q_4.

Three capacitances are contemplated. The first C_1, counts for the gate capacitance of Q_2 plus the junction capacitances of the drains of Q_{1b} and Q_{3b}. The second encompasses the output load C_2 plus the parasitic junction capacitances of the drains of Q_2 and Q_4. The third is the Miller capacitance C_m, bridging the input and output nodes of the second stage. The role of this capacitor is explained hereafter.

8.2.2 Pole Splitting

Table 8.1 lists the transconductances and capacitances of the Miller Op. Amp considered for the analysis that follows. The gain-bandwidth product is supposed to be equal to 50 MHz.

The poles associated to the high impedance nodes 1 and 2, respectively g_{d1}/C_1 and g_{d2}/C_2, determine cut-off frequencies respectively equal to 1.99 and 3.06 MHz. The amplifier is thus clearly a second order system since both poles lie well below the transition frequency. The purpose of the Miller capacitance is to split these poles apart, pushing one beyond f_T, the other to much lower frequencies. This turns the Op. Amp into a first order or dominant pole unconditionally stability circuit. The way C_m transforms the two poles into a very low frequency pole and a high frequency pole is reviewed briefly hereafter. To illustrate the mechanism, we analyze the frequency response of the Miller Op. Amp from DC to high frequency.

The DC gain is obtained by multiplying the DC gain A_1 of the first stage by the DC gain A_2 of the second stage. These are respectively equal to 41 and 39 dB making the overall gain equal to 80 dB:

$$A_o = A_1 \cdot A_2 = \frac{g_{m1}}{g_{d1}} \cdot \frac{g_{m2}}{g_{d2}} \qquad (8.1)$$

Now let us increase progressively the frequency. The first capacitance that is going to affect the performances of the amplifier is the Miller capacitance. The current flowing through C_m is much larger indeed than the current flowing through C_1 even though the magnitudes of the two capacitances are similar. The reason is that the voltage difference across C_m is an enlarged replica of the voltage across C_1 equal to $(1 - A_2)v$. The impact on node 1 of the Miller capacitance can be emulated consequently by means of a grounded capacitance equal to $(1 - A_2)$ times C_m. The admittance y of node 1 is given consequently by:

$$y = j\omega \, C_m \, (1 - A_2) \approx j\omega \, C_m \frac{g_{m2}}{g_{d2}} \qquad (8.2)$$

This is a huge capacitance that enhances considerably the time constant associated to node 1 and gives raise to a the low frequency pole that can be approximated by the expression:

$$\omega_1 = \frac{g_{d1}}{A_2 C_m} = \frac{g_{d1}}{C_m} \cdot \frac{g_{d2}}{g_{m2}} \qquad (8.3)$$

According to the data listed in Table 8.1, the pole associated to node 1 is positioned at 5 kHz. The consecutive break affecting the gain A_1 is visible in Fig. 8.3, Beyond 5 kHz, and as long as the gain of the second stage is real, the magnitude of the gain of the first stage decreases steadily at −20 dB/decade. Things change however when the cut-off frequency of the second stage is reached. The angular cut-off frequency of this stage is given by the expression below for transistor Q_2 is feeding not only C_2 but also C_m. The left terminal of the Miller capacitance connected to input gate of Q_2, may be assimilated indeed to the virtual ground of the sub-Op Amp represented by the second stage.

$$\omega_2 = \frac{g_{d2}}{C_2 + C_m} \qquad (8.4)$$

Beyond ω_2, the gain of the second stage becomes imaginary so that the transfer function can be approximated by the expression:

$$A_2 \approx -\frac{g_{m2}/(C_2 + C_m)}{j\omega} \qquad (8.5)$$

8.2 The Miller Op. Amp.

The 90° phase lag associated to A_2 modifies drastically the load on the first stage represented by the Miller capacitance for the nature of the latter changes from capacitive to resistive. Indeed, we must consider now that y is given by:

$$y = j\omega C_m \left(1 + \frac{\frac{g_{m2}}{C_2+C_m}}{j\omega}\right) \approx g_{m2} \frac{C_m}{C_2 + C_m} \qquad (8.6)$$

The load on node 1 boils down now to a small resistance equal to $1.6\,\mathrm{k}\Omega$. Now that A_1 is loaded by a resistance, the gain of the first stage remains constant. In fact, the pole of the second stage gives birth to a zero as far as the first stage. Both cancel out exactly so that the overall frequency response of the Op Amp ignores what is happening at node 1 and continues to decay steadily as shown in Fig. 8.3. But this holds true only as long as the gain of the second stage is large enough to sustain the resemblance with a sub-Op. Amp. When the gain of the second stage is no more than a few dB's, the approximation isn't correct anymore of course. This is what happens at high frequency when the capacitances overwhelm the conductances. The transfer function of the Op. Amp boils down then to:

$$A = \frac{g_{m1}}{j\omega C_m} \cdot \frac{g_{m2} - j\omega C_m}{g_{m2} + j\omega(C_1 + C_2 + C_1 C_2 / C_m)} \qquad (8.7)$$

While the factor in front of the above expression takes care of the $-20\,\mathrm{dB/decade}$ roll-off, the second reveals what happens near and beyond the transition frequency.

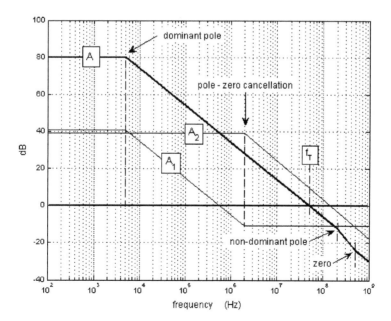

Fig. 8.3 Asymptotic frequency response of the Miller Op. Amp.

A pole and a zero are acknowledged in this region. The zero lies in the right part of the complex plane, the pole in the left part.

The R.H.P. (for Right Half Plane) zero witnesses actually the fact that the Miller capacitance bypasses Q_2 at very high frequency. The signal from the first stage reaches the output terminal directly through C_m wiping out the 180° phase shift inherent to the second stage. Unfortunately, this introduces a global 90° excess phase lag, which reduce the phase margin. Cumulated with the 180° phase shift of the dominant and the non-dominant poles, the total phase lag amounts now to 270°. Stability requires consequently that the R.H.P zero and non-dominant pole be put to the right of the angular transition frequency. This implies that the gain-bandwidth product of the second stage must exceed ω_T in order to keep the resistive character of the impedance of node 1 near the transition frequency. The price one has to pay therefore is a low gain of the first stage in this region.

The gain bandwidth product of the Miller amplifier lies now for the hand. Since the Op Amp may be assimilated to a first order system, ω_T is equal to the product of the dominant pole times the DC gain. This leads to the well-known expression of the angular transition frequency:

$$\omega_T = \frac{g_{m1}}{C_m} \tag{8.8}$$

8.2.3 The Impact of the Current Mirror

Figure 8.4 shows the open-loop frequency response of the Miller Op Amp derived the symbolic expression listed under Eq. 8.10 and compares the result to the data displayed by Fig. 8.3.

The frequency responses are almost identical except far beyond the transition frequency. The explanation is due to the current mirror. In the analysis above, we assumed that the current entering node 1 is the algebraic sum of the drain currents delivered by Q_{1a} and Q_{1b}. This is a simplification for it ignores the time lag associated to the current mirror. The AC current feeding node 1 consists indeed of two distinct currents, current from Q_{1b} reaching node 1 directly, and current from Q_{1a} transiting through the current mirror. The voltage drop across the diode connected transistor Q_{3a} controls the current delivered by the mirror. This introduces a time constant that depends on the conductance g_{m3} of Q_{3a} and the parasitic gate capacitances of transistors Q_{3a} and Q_{3b} plus the parasitic junction capacitances of Q_{3a} and Q_{1a}. This time constant is much smaller than that of node 1 for the conductance g_{m3} of the diode is much larger than g_{d3} whereas the parasitic capacitance C_3 does not differ substantially from that of node 1 (remember the ratio g_{m3}/g_{d3} represents the intrinsic gain of Q_3).

The impact can be accounted for by multiplying g_{m1} by a correction factor:

$$g_{m1} := g_{m1} \frac{g_{m3} + j\omega\, C_3/2}{g_{m3} + j\omega\, C_3} \tag{8.9}$$

8.2 The Miller Op. Amp.

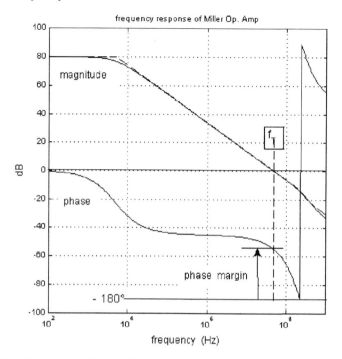

Fig. 8.4 *Plain lines* represent the magnitude and phase of the Miller Op. Amp frequency response derived from the symbolic equations listed under Eq. 8.10. The magnitude is compared to the asymptotic counterpart represented by means of *dashed lines*

The current mirror introduces thus a pole and a zero one octave beyond, in other words a doublet. In the example, the frequencies corresponding to the pole and zero are respectively 144 and 288 MHz. Since they are almost three times larger than the transition frequency, the doublet has practically no influence on the phase margin.

8.2.4 Poles and Zeros

The pole splitting effect is clearly visible in the plot of Fig. 8.5, which shows the trajectories of the poles and zeros of the Op. Amp when C_m varies from a very small to a very large value (singularities can be obtained by means of the *roots* MATLAB instruction). The numerator and the denominator of the transfer function obtained be means of a symbolic simulator are listed hereunder:

% numerator
N2 = −2*Cm*C3*gm1;
N1 = −2*Cm*gm3*gm1 + C3*gm2*gm1;

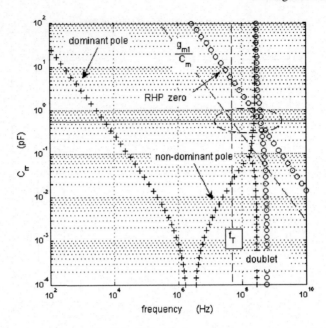

Fig. 8.5 Pole and zeros versus C_m of the Miller Op Amp with the current mirror doublet

```
N0 = 2*gm2*gm3*gm1;
num = [N2 N1 N0];

% denominator
D3 = 2*Cm*C1*C3 + 2*Cm*C2*C3 + 2*C1*C2*C3;
D2 = 2*Cm*C3*gd1 + 2*C2*C3*gd1 + 2*Cm*C3*gd2 + 2*C1*C3*gd2 ···
     +Cm*C3*gm2 + 2*Cm*C1*gm3 + 2*Cm*C2*gm3 + 2*C1*C2*gm3;
D1 = 2*C3*gd1*gd2 + 2*Cm*gd1*gm3 + 2*C2*gd1*gm3 + ···
     +2*Cm*gd2*gm3 + 2*C1*gd2*gm3 + 2*Cm*gm2*gm3;
D0 = 2*gd1*gd2*gm3;
den = [D3 D2 D1 D0];
```
(8.10)

As long as the Miller capacitance is negligible, the poles associated to nodes 1 and 2 are clearly distinguishable at the bottom of the figure together with the high frequency doublet. Pole splitting takes place when C_m increases. The dominant pole moves left. The other goes right until it merges with the pole of the doublet forming a complex conjugate pair of which only the real part is visible. As C_m further increases, the Right Half Plane (RHP) zero enters the plot, slowly overruling high frequency poles and zeros. The optimal combination lies naturally in the region marked by the ellipse where the magnitude of C_m is just large enough to turn the Op. Amp into a first order system. The plain horizontal line in the middle represents the capacitance reported in Table 8.1. The dominant pole lies then at 5 kHz while the aggregate consisting of a complex conjugate pole plus a left and a right-sided zero lies beyond ω_T. When C_m goes above the horizontal line, the transition frequency follows the dashed curve predicted by Eq. 8.8, paralleling the RHP zero locus leaving other singularities far away. The dominant pole and the R.H.P zero control the phase lag almost exclusively. Though clearly unconditionally stable, the Op. Amp displays a gain-bandwidth product that is severely impaired by a needlessly too large C_m. Below the horizontal line, the validity of the dominant pole approximation ceases while the influence of the other singularities increases. The order of the system increases causing a more rapid drop of the angular transition frequency than what is suggested by the dashed line.

A well-known technique allowing to get rid of the phase lag associated to the RHP zero consists in putting a resistor in series with the Miller capacitance. If this resistance is equal to the reciprocal of the transconductance of the second stage, the zero is relegated to infinity. Larger resistances return the zero into the left half complex plane offering the possibility to perform eventually pole-zero cancellations. The method is not for free for it generates another far end pole threatening the phase margin. When the resistor is properly calibrated however, the overall gain-bandwidth product can be enhanced by a factor nearly equal to two.

8.3 Sizing the Miller Operational Amplifier (MATLAB OpAmp.m)

So far for the analysis, let us focus now on sizing. The output voltage of the Miller Op. Amp is supposed to be equal to $V_{DD}/2$ while the input terminals connected to a symmetrical small signal source is centered half the power supply. We consider only the right-sided P-channel transistor Q_{1b} for its source potential may be assimilated to an artificial ground owing to symmetry. Though the input signal is halved, the transconductance of Q_{1b} needs not to be divided by two for the current mirror doubles the small signal current. V_{S1} and the gate voltage of Q_2 are not fixed yet for they depend on currents and aspect ratios unknown so far. The gate voltage of Q_4 and Q_5 are left open to keep a degree of freedom.

We mentioned in the beginning that gain-bandwidth and unconditional stability determine the *specifications* of the Op. Amp whereas power consumption and area are *attributes*. The gain-bandwidth product is given by Eq. 8.8. The right half plane zero and non-dominant pole controlling the phase margin, are extracted from Eq. 8.7[1]:

$$\omega_Z = \frac{g_{m2}}{C_m} \qquad (8.11)$$

$$\omega_{NDP} = \frac{g_{m2}}{C_m} \cdot \frac{C_m^2}{C_m(C_1 + C_2) + C_1 C_2} \qquad (8.12)$$

---

[1] Singularities like the doublet associated with the current mirror are omitted for they lie beyond the gain-bandwidth product. Their impact is considered later.

We must now choose suitable ω_Z/ω_T and ω_{NDP}/ω_T ratios. We call these respectively Z and NDP. Generally, the zero is put one decade beyond the gain-bandwidth product and the non-dominant pole somewhere between. A Z equal to ten and NDP equal to four yield a phase margin around 60° to 70°. The transconductances of Q_1 and Q_2 can be extracted then from Eqs. 8.8, 8.11 and 8.12:

$$g_{m1} = \omega_T C_m \qquad (8.13)$$
$$g_{m2} = \omega_Z C_m = Z g_{m1} \qquad (8.14)$$

While ω_T and Z and NDP are known a priori, the Miller capacitance C_m is unknown. If the parasitic capacitances of nodes 1 and 2 were available, C_m could be extracted from the inverted Eq. 8.12:

$$C_m = \frac{NDP}{Z} \left(C_1 + C_2 + \sqrt{(C_1 + C_2)^2 + 4 C_1 C_2 \frac{Z}{NDP}} \right) \qquad (8.15)$$

The capacitance of node 1 requires knowing the gate-to-source capacitance of transistor Q_2 plus the parasitic junction capacitances of Q_{3b} and Q_{1b}. Similarly, C_2 requires knowing the parasitic capacitance paralleling the output load capacitance. None of these are known so far for the widths are not fixed yet. What is possible however is to estimate a likely value of C_m, derive from this guess the corresponding transconductances g_{m1} and g_{m2}, then find currents and aspect ratios by means of the g_m/I_D methodology and there from derive approximated parasitic capacitances from the transistor sizes. The new Miller capacitance extracted from Eq. 8.15 can be reutilized to redo the same calculations until a stable C_m is obtained. A few additional constraints should be fixed in the same time. The gate-to-source voltage of Q_{3b} and the gate-to-source voltage V_{GS4} are not known so far. For what concerns the drain voltage of Q_3, it should be a replica of the drain voltage of Q_{3a} to avoid a systematical input offset of the differential stage. Consequently, the gate-to-source voltage of Q_{3b} must be equal to the gate-to-source voltage of Q_2. For what concerns transistor Q_4, keep in mind that this P-channel device is driving 90% of the total current. It might be very large. One may be better off fixing W_4 instead of V_{GS4} for this offers a direct control over the area occupied by Q_4.

8.3.1 Sizing a Low-voltage Miller Op. Amp.

We implement in this section a low-voltage Op. Amp (1.2 V) loaded by a 3 pF capacitor that achieves a gain-bandwidth product of 20 MHz. Small gate lengths are not required, for the transition frequency is low. We choose larger gate lengths to enhance gain.

	Q_1	Q_2	Q_3	Q_4	Q_5
L (µm)	1.0	0.5	1.0	0.5	1.0

8.3 Sizing the Miller Operational Amplifier (MATLAB OpAmp.m)

Sizing is performed now in a 2D sizing space, one axis per stage. Several issues are possible. The variables may be the g_m/I_D's of Q_1 and Q_2 or suitable ranges of q_{F1} and q_{F2}. We opt for the second and set up q_{F1} and q_{F2} matrices encompassing moderate inversion to optimize power consumption without the risk to end up with oversized transistors.

```
X = logspace(-1.,0,20);            % qF1 horiz.
Y = logspace(-1.,0,30)';           % qF2 vert.
[qF1,qF2] = meshgrid(X,Y);
```

Next, we evaluate the parameters of every transistor over the entire sizing space. For Q_2 the appraisal is straightforward since the source and drain voltages V_{S2} and V_{D2} are respectively equal to 0 to $V_{DD}/2$:

```
vds2  = round(40*VDS2 + 1);
vs2   = round(10*VS2 + 1);
n2    = nN(vds2,vs2,lg2);
VTo2  = VToN(vds2,vs2,lg2);
ISuo2 = ISuoN(vds2,vs2,lg2);
```

Knowing the parameters, we evaluate the unary drain current I_{Du2} and gate voltage V_{GS2} matrices of Q_2:

```
i2   = qF2.^2 + qF2;
IDu2 = i2*ISuo2;
VPS2 = UT*(2*(qF2 - 1) + log(qF2));
VGS2 = n2*VPS2 + VTo2;
```

Next consider the parameters of Q_{3b}. The source of Q_{3b} is grounded while the drain voltage is fixed by V_{GS2}. The V_{GS2} matrix does comply with the nominal entries of the parameter matrices. Consequently, they do not give access to parameter look-up tables. Every parameter must be interpolated. For I_{Du3}, the evaluation follows the same lines as above.

```
vs3   = round(10*VS3 + 1);
n3    = interp1(U,nN(:,vs3,lg3),VGS2,'cubic');
VTo3  = interp1(U,VToN(:,vs3,lg3),VGS2,'cubic');
ISuo3 =
interp1(U,ISuoN(:,vs3,lg3),VGS2,'cubic');
VPS3  = (VGS2 - VTo3)./n3;
qF3   = invq(VPS3/UT);
i3    = qF3.^2 + qF3;
IDu3  = i3.*ISuo3;
```

Things are a bit more complicated for Q_1. All we know so far is that its gate voltage is equal to $V_{DD}/2$ and the drain voltage a function of V_{GS2}.[2] Though the source

---

[2] For P-channel transistors, voltages are defined with respect to V_{DD}.

voltage is unknown for it depends on currents and sizes not fixed yet, V_{S1} can be anticipated more or less for sensible gate-to-source voltages lie around 0.4 V. We consider a likely V_{S1} even if the source voltage must be corrected after sizing. For the rest, the procedure is the same as with Q_2.

```
vs1   = round(10*VS1 + 1);
VDS1  = VDD - VS1 - VGS2;
n1    = interp1(U,nP(:,vs1,lg1),VDS1,'cubic');
VTo1  = interp1(U,VToP(:,vs1,lg1),VDS1,'cubic');
ISuo1 =
interp1(U,ISuoP(:,vs1,lg1),VDS1,'cubic');
i1    = qF1.^2 + qF1;
IDu1  = i1.*ISuo1;
VPS1  = UT*(2*(qF1 - 1) + log(qF1));
VGS1  = n1.*VPS1 + VTo1;
```

The source and the drain voltages of Q_4 are respectively 0 V and $V_{DD}/2$. Instead of choosing V_{GS4}, we make W_4 equal to W_2 to keep control over the size of Q_4 as mentioned earlier. Knowing W_4 we can evaluate the unary drain current, there from the normalized drain current i_4 and q_{F4} and get the gate voltage V_{GS4}. Notice that q_{F4} is necessarily larger than q_{F2}, which is a desirable feature as far as sensitivity of the current sources feeding the Op. Amp.

```
vds4  = round(40*VDS4 + 1);
vs4   = round(10*VS4 + 1);
n4    = nP(vds4,vs4,lg4);
VTo4  = VToP(vds4,vs4,lg4);
ISuo4 = ISuoP(vds4,vs4,lg4);
WsL4  = W4/LL(lg4);
IDu4  = ID2./WsL4;
i4    = IDu4/ISuo4;
qF4   = .5*(sqrt(1 + 4*i4) - 1);
VPS4  = UT*(2*(qF4 - 1) + log(qF4));
VGS4  = n4.*VPS4 + VTo4;
```

The evaluation of the parameters and unary drain current of Q_5 is straightforward and follows the same lines as with Q_2. The source, drain and gate voltages are respectively 0 V, V_{S1} and V_{GS4}.

We now review briefly the *specifications* list before launching the sizing algorithm:

1. The transconductance g_{m1} given by Eq. 8.13 requires to know the Miller capacitance. Since this is not the case, we choose a plausible C_m, for example half the output capacitance (the choice is not critical). Running the sizing algorithm a few times yields the actual Miller capacitance.
2. The tranconductance g_{m2} is extracted from Eq. 8.14. Z controls the position of the R.H.P with respect to the gain-bandwidth product, it is chosen equal to ten.

3. The position of the non-dominant pole with respect to the gain-bandwidth product controls the phase margin together with the R.H.P zero. A phase margin of 60° to 70° requires an N.D.P of four.[3]

The excerpt below shows the actual sizing algorithm. The drain current and width matrices of Q_1 and Q_2 are evaluated taking advantage of the g_m/I_D methodology. W_1 and W_2 and the widths of Q_3 and Q_4 are combined in order to evaluate the parasitic capacitances of nodes 1 and 2 where from we derive a new C_m according to Eq. 8.15.

```
Cm = .5*C;
for k = 1:10,
    GM1 = 2*pi*fT*Cm;
    ID1 = n1*UT.*(1 + qF1).*GM1;
    WsL1 = ID1./IDu1;
    W1 = WsL1*LL(lg1);
    GM2 = Z*GM1;
    ID2 = n2*UT.*(1 + qF2).*GM2;
    WsL2 = ID2./IDu2;
    W2 = WsL2*LL(lg2);
    WsL3 = ID1./IDu3;
    W3 = WsL3*LL(lg3);
    W4 = W2;
    y1 = jctCap(LL(lg1),W1,R,VDD-VGS2);
    y3 = jctCap(LL(lg3),W3,R,VGS2);
    C1 = y1(:,:,1) + y3(:,:,1) + W2.*CGS2u;
    y2 = jctCap(LL(lg2),W2,R,VDS2);
    y4 = jctCap(LL(lg4),W4,R,VDD-VDS2);
    C2 = C + y2(:,:,1) + y4(:,:,1);
    Cm = 0.5*NDP/Z*(C1 + C2 + sqrt((C1+C2).^2 + C1.*C2*4*Z/NDP));
end
```

The contribution of the gate-to-source capacitance of Q_2 to node 1 is derived in the file above from the *global* semi-empirical capacitance C_{GS}. A representation of the gate-to-source capacitance based on the compact model is described in Annex 4.

This brings the first part of the sizing procedure to an end. Matrices representing currents, widths of all the transistors are now available.[4] Every point in the sizing space points towards currents or widths of distinct Op. Amps fulfilling the same gain-bandwidth and phase margin *specifications*. To compare their performances, we have to weight now *attributes*, for instance power consumption against 'active'

---

[3] The positions of the poles and zeros beyond the angular transition frequency play a major role. A change of NDP can cause substantial modifications. A drop from 4 to 3 may cause ringing, while an increase from 4 to 5 enhances the power consumption by nearly 25% without improving the step function response.

[4] Rows depend on q_{F1}, columns on q_{F2}.

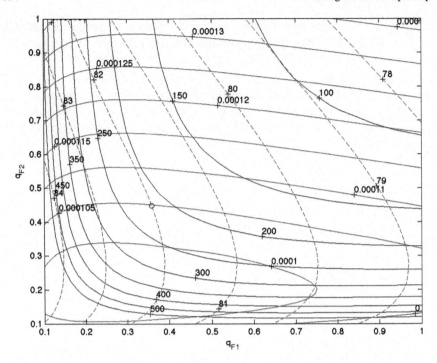

Fig. 8.6 Constant D.C current (*plain elliptic*), constant active area (*plain hyperbolic*) and constant gain (*dashed*) contours versus first and second stage q_F's

area.[5] One can make use of 3D representations to 'visualize' the performances or consider contours plots like in the constant power and area contours displayed in Fig. 8.6. The hyperbolic-shaped contours delineate 'active areas', while elliptic contours determine the total D.C current. Contours show clearly the pros and cons of the choices we can make. Small q_F's for instance, mean less current and larger transistors. If q_F gets too small, parasitics get so large that the total current starts to grow again like in the I.G.S. Opposed, if we increase the D.C current by 10%, the 'active' area can be divided by a factor of four.

Constant gain contours derived from the semi-empirical data are displayed in the same plot (dashed lines). As currents decrease, the gain increases of course owing to larger g_m/I_D's. Though clearly visible in the upper part of the figure, the trend seems to revert slightly in the lower part suggesting that the second stage should not go too deep in moderate inversion. When q_{F2} decreases, the gate-to-source voltage of Q_2 tends to lessen indeed. Since the drain and gate voltages of the current mirror's transistors follow the same trend, widths must grow to sustain the current delivered by the differential stage. Consequently, the conductance of Q_{3b} increases and this has a harmful effect on the gain of the first stage.

---

[5] By 'active area', we mean the sum of the areas occupied by the gate and junctions of every transistor.

8.3 Sizing the Miller Operational Amplifier (MATLAB OpAmp.m)

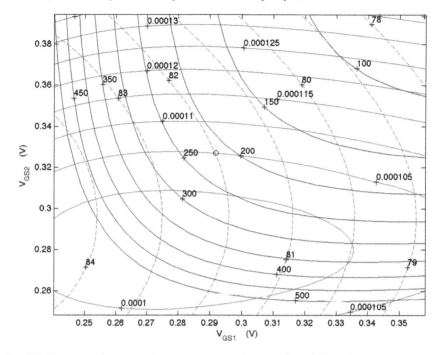

Fig. 8.7 Representation versus the gate-to-source voltages of Q_1 and Q_2 of the data displayed in Fig. 8.6

Table 8.2 Gate voltages, currents, widths and transconductances of the 20 MHz gain-bandwidth Miller Op. Amp that corresponds to the circle displayed in Figs. 8.6 and 8.7

	Q_1	Q_2	Q_3	Q_4	Q_5
q_F	**0.35**	**0.45**	0.57	1.54	
V_{GS}(V)	0.292	0.327	0.327	0.392	0.392
I(mA)	0.0069	0.0912	0.0069	0.0912	0.0138
W(μm)	66.0	47.9	5.61	47.9	24.0
g_m(mS)	0.176	1.76	0.125		

Another presentation than the one shown in Fig. 8.6 may be desirable for normalized mobile carrier densities are not widespread design parameters. Figure 8.7 displays the same data, versus gate-to-source voltages of Q_1 and Q_2. The passage from one representation to the other is easy for the connection from q_F to V_{GS} via V_P is straightforward.

Let us choose an Op. Amp, for instance the one marked by a circle in the two last figures. The Op. Amp has a gain equal to 82 dB (43.95 for the first stage and 38.06 dB for the second). The D.C current is equal to 105 μA and the active area is comprised between 200 and 250 μm². With the recommended Miller capacitance of 1.41 pF, the phase margin is 70.2°. Table 8.2 lists the currents, widths, transconductances, gate-to-source voltages versus the q_F's of every transistor (the q_F coordinates of the circle are of printed in bold characters).

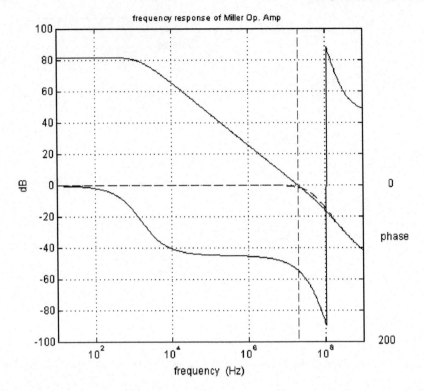

Fig. 8.8 Frequency response (magnitude and phase) of the Op. Amp marked by the circle in the two preceding figures. The *dashed line* represents the unity gain response

Going up along the 82 dB gain contour, the power consumption increases slightly while the area decreases a little before increasing again. Downwards, the supply current drops by a few % but the area grows rapidly. Along the constant 105 µA contour, the active area decreases before the opposite takes place while loosing gain. If area counts more than gain and D.C current, larger q_F's (or gate voltages) are recommended. The choice is a matter of ruling.

Figure 8.8 shows the frequency response of the open-loop (magnitude and phase) and unity-gain configurations (illustrated by means of the dashed line). The plot makes use of the transfer function given by Eq. 8.10 and the small signal parameters listed above. Consequently, the current mirror doublet ignored during the sizing process is acknowledged. As expected, the doublet does not impair the performances of the Op. Amp, but its impact is clearly visible in the poles-zeros display represented in Fig. 8.9. Like in Fig. 8.5, the non-dominant pole and the pole of the doublet merge and form a conjugated pair as they get closer. The merge takes place when C_m is nearing the plain horizontal line representing the recommended Miller capacitance. The configuration for the recommended Miller capacitance is the following:

8.3 Sizing the Miller Operational Amplifier (MATLAB OpAmp.m)

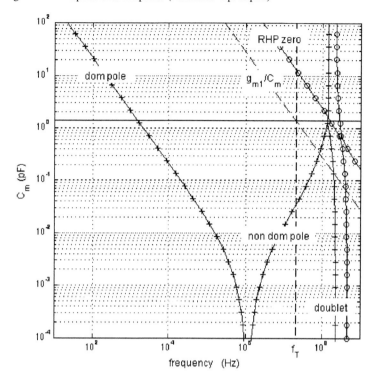

Fig. 8.9 Evolution of the zeros and poles as the Miller capacitance varies from a very small to a very large value

dom. pole :	$-1.55 \; 10^3$
complex conj. pole:	$(-1.31 +/- 0.24) \; 10^8$
doublet zero :	$-2.82 \; 10^8$
RHP zero :	$+1.59 \; 10^8$

The next figure illustrates the step function response of the Op. Amp. The quasi-exponential evolution in the right part shows that the combination of Z and NDP derived from the phase margin *specifications* is adequate notwithstanding the fact that the current mirror doublet has been ignored during sizing. Steady state conditions are reached after nearly 40 ns. Notice that the output voltage goes first briefly in the wrong direction. The high frequency bypass between node 1 and the output represented by the Miller capacitance explains the glitch. The signal outputted by the differential stage reaches the output before the second stage has the time to react. (Fig. 8.10).

More *attributes* can be incorporated of course, the 1/f noise generated by the N-channel current mirror for instance. Because the noise is proportional to the reciprocal of the gate area, it may seem tempting to increase the gate length and width

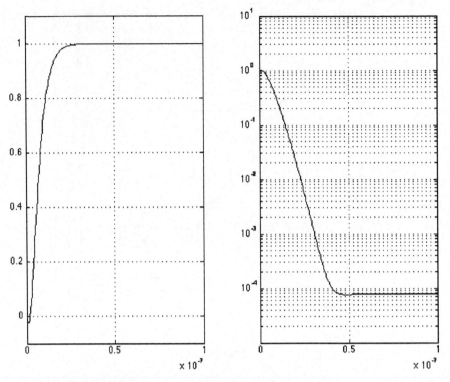

Fig. 8.10 The step function response of the unity gain Op. Amp is represented *left*. The *right* plot represents the difference between the input and output signals

of the current mirror transistors Q_3. Not only, the 1/f noise will lessen, but the gain will improve also owing to the lessening of the conductance of Q_{3b}. Suppose we extend the 1 μm gate lengths of the current mirror transistors to 4 μm. Figure 8.11 shows the modified area and the gain contours. The constant current contours don't change notably. Suppose we choose the Op. Amp marked by the circle. The gain increases by 4 dB's with respect to Fig. 8.7 while the total D.C current is kept unchanged. The active area reaches now 350 μm² while the currents and widths listed in Table 8.3 replace those of Table 8.2.

Figure 8.12 shows the impact of the larger parasitic capacitance of node 1 that is caused by the nearly 16 times larger area of Q_3. This moves the doublet closer to ω_T as is illustrated by the new set of poles and zeros:

dom. pole :	$-0.976 \, 10^3$
complex conj. pole:	$(-0.304 +/- 0.260) \, 10^8$
doublet zero :	$-0.344 \, 10^8$
RHP zero :	$+1.15 \, 10^8$

8.3 Sizing the Miller Operational Amplifier (MATLAB OpAmp.m)

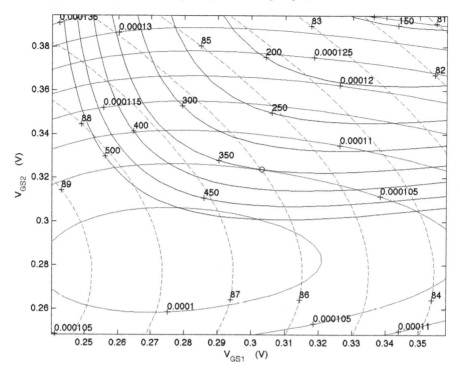

Fig. 8.11 Same as Fig. 8.7 when the gate length of Q_3 is increased from 1 to 4 μm in an attempt to reduce the 1/f noise generated by the N-channel current mirror

Table 8.3 Gate voltages, currents, widths and transconductances of the 20 MHz gain-bandwidth Miller Op. Amp corresponding to the circle displayed in Fig. 8.11

	Q_1	Q_2	Q_3	Q_4	Q_5
q_F	**0.44**	**0.43**	0.65	1.48	
V_{GS}(V)	0.303	0.324	0.327	0.387	0.387
I(mA)	0.0074	0.0903	0.0074	0.0903	0.0148
W(μm)	52.3	50.7	20.7	50.7	28.5
g_m(mS)	0.177	1.77	0.131		

The complex conjugate pole and the zero of the doublet are only a little more than one octave beyond ω_T. Though the phase margin is still 65° because the specifications did not change, the 'harmless' doublet causes a lot of ringing as shown in Fig. 8.12. The time to reach steady state conditions is almost multiplied by a factor two. Increasing the area of Q_3 beyond 1 μm is definitely not a good idea.

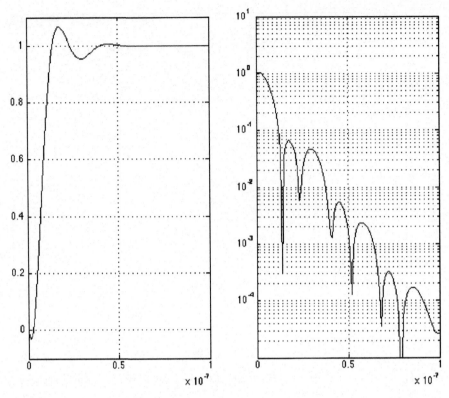

Fig. 8.12 The step function response of the unity gain Op. Amp when the size of transistors Q_3 is enhanced

8.3.2 Sizing a High-Frequency Low-Power Miller Op. Amp.

Let us now trade gain for speed aiming at a gain-bandwidth product of 200 MHz. The supply voltage, output capacitance, etc. don't change. The gate lengths are shortened of course: 160 and 130 μm for Q_1 and Q_2 respectively, 160 μm for Q_4 and 500 μm for Q_3 and Q_5. Moderate inversion implementations are still feasible.

Figure 8.13 shows the sizing space versus the V_{GS} axes. Transistor sizes and currents are reported in Table 8.4. The currents are nearly ten times larger than in the previous Op. Amp. The minimum supply current is around 1 mA. Gain is smaller of course. The active area doesn't change much. It is even smaller for shorter gate lengths help to achieve larger aspect ratios without enhancing necessarily widths. Suppose we select the Op. Amp marked by the circle at the crossing of the 1.1 mA and 61.5 dB contours. The gains of the first and second stages are respectively 35.7 and 25.8 dB, while the 'active' area is smaller than 100 μm². Notwithstanding the larger gain-bandwidth product, we can meet the *specifications* with Q_1 and Q_2 still being in moderate inversion.

8.3 Sizing the Miller Operational Amplifier (MATLAB OpAmp.m)

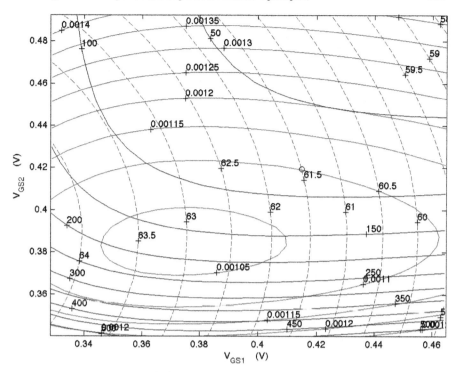

Fig. 8.13 Sizing space of the 1.2 V Miller Op. Amp achieving a gain-bandwidth product of 200 MHz (same 3 pF load as in previous figures)

Table 8.4 Gate voltages, currents, widths and transconductances of a 200 MHz gain-bandwidth Miller Op. Amp corresponding to the circle displayed in Fig. 8.12

	Q_1	Q_2	Q_3	Q_4	Q_5
q_F	0.55	0.43	1.23	1.44	
V_{GS}(V)	0.415	0.430	0.430	0.474	0.474
I(mA)	0.089	0.922	0.089	0.922	0.178
W(μm)	54.3	123.6	11.3	123.6	29.2
g_m(mS)	1.74	17.4	1.12		

The recommended Miller capacitance is equal to 1.38 pF and the phase margin equal to 70.6°. The step function response is similar to that of Fig. 8.10 with the exception of the horizontal scale, which is divided of course by ten. The poles and zeros are:

$$
\begin{array}{ll}
\text{dom pole:} & -1.59\ 10^5 \\
\text{complex conj. pole:} & (-1.02 +/- 0.555\,i)\ 10^9 \\
\text{doublet zero:} & -2.13\ 10^9 \\
\text{RHP zero:} & 1.60\ 10^9
\end{array}
$$

The Op. Amp reaches steady state conditions after nearly 5 ns. Since the slew-rate $2I_{D1}/C_m$ is equal to $128\,\text{V}/\upmu\text{s}$, the largest output voltage swing tolerated after 5 ns is equal to 0.64 V (nearly half the power supply). Notice that when the slew-rate is specified instead of the gain-bandwidth product, sizing can be performed along similar lines. The current in the first stage is known for it is fixed by the load capacitance and the slewing rate. The width of Q_1 is evaluated the same way and nothing changes for the rest. The drain current I_{D1} is automatically updated after each run.

8.4 Conclusion

The Miller Op. Amp discussed in this chapter broadens the conclusions made earlier about the Intrinsic Gate Stage. The methodology is similar but a stronger *specification-attribute* dichotomy is needed to separate clearly specifications from optimization objectives. The first determine the dimensions of the sizing space. The second delineate optimal areas within the optimization space. The method doesn't select 'the' ideal implementation but orients choices. The intersection of a constant current contour with a constant 'active area' contour for instance tells us that two implementations with the same power consumption and silicon real estate are possible. Shifting the 'active area' contour in the up/right direction until the two points merge paves the way to the smallest 'active area' possible for a given supply current. Repeating the experiment with different currents, connects power consumption to minimal area implementations.

A thorough understanding of the Op. Amp's behavior is recommended of course to separate first order from second order objectives. The simplifications bring about the need for checking results by means of high level tools. The merit of the g_m/I_D method is that the designer is guided towards nearly optimal implementations keeping fine tuning for advanced simulation tools.

Annex 1
How to Utilize the Data available under 'extras.springer.com'

The data provided under 'extras.springer.com' consist of look-up tables listing the semi-empirical data and E.K.V. parameters of the N- and P-channel devices considered throughout the book. These are *global variables* that must be declared before undertaking any other action.

A1.1 Global Variables

Thz *Glob.m* file residing in the *0 start* directory must be run before any other file in order to declare the *global variables* (this must be done once when starting). The *global variables* encompass the arrays listed hereunder:

Semi-empirical Global Variables – Courtesy of IMEC

1. Drain currents (W = 10 μm)
       IDRAINn/p Drain currents[1] I_D
2. Transconductance (W = 10 μm)
       GMn/p   The gate transconductance g_m
       GMBn/p   The back-gate transconductance g_{mb}
       GDSn/p   The drain conductance g_{ds}
3. Intrinsic capacitances (W = 10 μm)
       CGGn/p   The gate capacitance C_{gg}
       CGSn/p   The gate-to-source capacitance C_{gs}
       CGDn/p   The gate-to-drain capacitance C_{gd}
       CGBn/p   The gate-to-substrate capacitance C_{gb}
       CSSn/p   The source capacitance (com-gate) C_{ss}
4. Gate lengths (μm)
       LL   Available gate lengths

and

---

[1] The lower case letter ending each array refers to N- or P-channel transistors.

Compact Model Global Parameters (W/L = 1)

nN/P	Slope factor	n
VToN/P	Threshold voltage	V_{To}
ISuoN/P	Unary specific current	I_{Suo}
ThN/P	Mobility degradation factor	
PolyN/P	Theta polynomial	

A1.2 An Example Making Use of the 'Semi-empirical' Data: The Evaluation of Drain Currents and g_m/I_D Ratio Matrices (MATLAB A12.m)

Once Glob.m run, files can access **global variables**. The name of the **global variables** put to use must be listed on top of the files. For instance, a file making use of N-channel drain currents must begin with:

$$\textit{global IDRAINn}... \tag{A1.1}$$

IDRAINn (like any other **global variable**, transconductance or capacitance) consists of a '9 by 49 by 49 by 9 4D array that can be accessed by means of subscripts specifying addresses: lg for the 9 available gate lengths, vgs for the 49 gate-to-source voltages, vds for the 49 drain-to-source voltages and vs for the 9 source-to-substrate voltages.

The available gate lengths are listed under the **global variable** LL:

$$LL = [0.10 \quad 0.11 \quad 0.12 \quad 0.13 \quad 0.14 \quad 0.16 \quad 0.50 \quad 1.00 \quad 4.00] \ \mu m \tag{A1.2}$$

The available gate-to-source V_{GS}, drain-to-source V_{DS} and source-to-substrate voltages V_S are:

Gate-to-source voltages	0 : 0.025 : 1.200(V)
Drain-to-source voltages	0 : 0.025 : 1.200(V)
Substrate voltages	0 : 0.100 : 0.800(V)

(A1.3)

Voltages can be translated into addresses[2]:

$$vgs = round\,(40 * VGS + 1) \tag{A1.4}$$
$$vds = round\,(40 * VDS + 1) \tag{A1.5}$$
$$vs = round\,(10 * VS + 1) \tag{A1.6}$$

To go from addresses to voltages, we make use of:

---

[2] The optional **round** instruction is recommended to avoid eventual non-integer subscripts resulting from arithmetic calculations.

A1.2 An Example Making Use of the 'Semi-empirical' Data

$$VGS = 0.025 * (vgs - 1) \quad (A1.7)$$
$$VDS = 0.025 * (vds - 1) \quad (A1.8)$$
$$VS = 0.1 * (vs - 1) \quad (A1.9)$$

Consider an example: suppose we want to construct the drain current matrix of a 100 nm (lg = 1) grounded source transistor (vs = 1) whose V_{GS} is swept across the full range of gate voltages while the drain voltage varies from 0.6 to 1.2 V in steps 0.2 V wide (VDS = 0.6: 0.2: 1.2). For vgs, a colon suffices since we consider all possible V_{GS}'s. For vds, according to A1.5:

$$vds = round(40 * VDS + 1);$$

The size of the resulting drain currents array, named ID, is [1 49 4 1].

$$ID = IDRAINn(lg, :, vds, vs)$$

To turn ID into a 49 rows and 4 columns matrix, one makes use of ***global variable the squeeze*** instruction:

$$ID = \boldsymbol{squeeze}(ID);$$

The file below computes the derivative of $\boldsymbol{log}(I_D)$ with respect to V_G to generate the g_m/I_D matrix and plot the result versus the gate voltage. The derivative takes advantage of the ***diff*** instruction. Since ***diff*** instructions are carried out vertically, the drain current matrix must be organized along gate controlled rows and drain controlled columns.

```
1 % test
2 clear
3 clf
4
5 global IDRAINn
6
7 lg = 1;
8 vs = 1;
9 VGS = (0:.025: 1.2)';  z = length(VGS);
10 VDS =.6:.2: 1.2;  vds = round(40*VDS + 1);
11 ID = squeeze(IDRAINn(lg,:,vds,vs));  size(ID)
12
13 gmID1 = diff(log(ID))/.025;
14 U =.5* (VGS(1:z-1) + VGS(2:z));
15 [X,Y] = meshgrid(VDS,U);
16 gmID = interp2(X,Y,gmID1,VDS,VGS);
17
18 plot(VGS,gmID,'k');
19 xlabel('V_G_S (V)'); ylabel('gm/ID (1/V)');
```

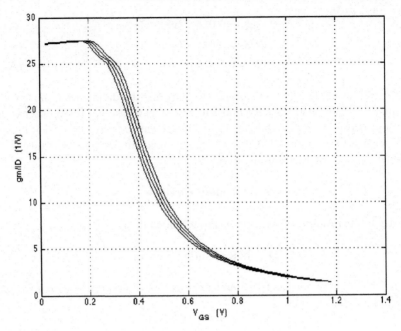

Fig. A1.1 The gm/ID curves obtained after running the file above

Care is needed regarding the size of gmID1. Owing to the differentiation, the number of rows of gmID1 is one step shorter than those of ID matrix. To get a g_m/I_D matrix with the same dimensions, the number of rows must be incremented by one unit. Resizing gmID1 in the vertical direction is done by means of the ***interp2*** instruction of line 16. We calculate therefore the X and Y matrix-coordinates of gmID1. This is done by means of the ***meshgrid*** instruction of line 15, which requires the pseudo-gate voltage U of gmID1 first. Figure A1.1 shows the final g_m/I_D curves. Notice that the same method can be put to use in order to calculate g_d/I_D when the drain current matrix is transposed before the ***diff*** instruction is performed.

A1.3 An Example Making Use of the E.K.V Global Variables: The Elaboration of an ID(VGS) Characteristic (Matlab A13.m)

The ***global variables*** nN/P, VToN/P and ISuoN/P, respectively the compact model slope factor, threshold voltage and unary specific current, of the compact model consist of '49 by 9 by 9' 3D arrays. These can be accessed by means of subscripts specifying vds, vs and lg, the same as with the 'semi-empirical' data. The model ignore V_{GS} by definition.

A1.3 An Example Making Use of the E.K.V Global Variables

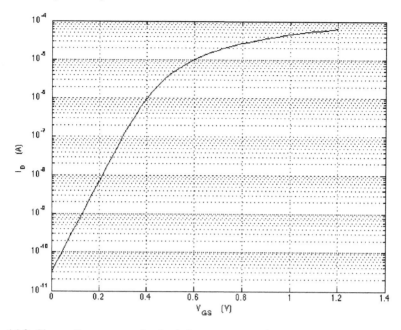

Fig. A1.2 The semilog representation the drain current versus the gate-to-source voltage resulting from the file above

The *global variables* PolyN/P and ThN/P are 4D arrays allowing to calculate the mobility degradation factor. The three first subscripts of both variables are the same as above. The fourth subscript of PolyN/P is always a colon. PolyN/P displays the coefficients ordered in descending powers of the mobility degradation polynomial. The number of coefficients is 5 (order 4 polynomial) and the argument of the polynomial the normalized drain current. The second global variable ThN/P makes use vds, vs, lg while the fourth subscript vgs calculates the degradation factor along the same lines as the polynomial representation.

The file below shows an example. The transistor is the same as above but the drain voltage is now constant and equal to 0.6 V. The slope factor n, the threshold voltage VTo and the unary specific current ISuo are scalars. A squeeze instruction is needed in order to turn the coefficients of the mobility degradation polynomial in to a vector. The calculation of the drain current is straightforward and follows the steps presented in Chapter 5. The result is shown in Fig. A1.2.

```
global nN VToN ISuoN PolyN
% data ------------------
lg = 1;
vs = 1;
VDS = 0.6;
% compute ----------------
UT =.026;
```

```
vds = round(40*VDS + 1);
n = nN(vds,vs,lg);
VTo = VToN(vds,vs,lg);
ISuo = ISuoN(vds,vs,lg);
P = squeeze(PolyN(vds,vs,lg,:));
VGS = 0:.025: 1.2;
VP = (VGS - VTo)./n;
qF = invq(VP/UT);
qR = invq((VP - VDS)/UT);
i = qF.^2 + qF - qR.^2 - qR;
ID = i.*ISuo./polyval(P,i);
% plot ------------------
semilogy(VGS,ID); grid
xlabel('V_G_S (V)'); ylabel('I_D (A)');
```

Annex 2
The 'MATLAB' Toolbox

A series of dedicated functions enabling to run MATLAB files referenced throughout the book are accessible in the toolbox. *It is strongly recommended to make use of the set path facility before running any file that makes use of functions of the toolbox.* If not done, the functions will not be accessed.

A2.1 Charge Sheet Model Files

The files hereafter are intended to reproduce figures of Chapters 2 and 3 and to carry out 'software experiments'.

A2.1.1 The pMat(T,N,tox) *Function*

The **pMat** function puts together the ***technology vector*** *p* (or ***matrix***) needed to run C.S.M. functions like the **IDsh** function. The input data are scalars and/or equal lengths row vectors representing: T (the temperature in K), N (the doping concentration expressed in at.cm^{-3}) and t_{ox} (the oxide thickness in nm). A sign is associated to the doping concentration N to differentiate semiconductor types, positive for N-type substrates, negative for P-type. Else, the sign is ignored. The output of the **pMat** function consists of (a) column vector(s). The three first rows list ϕ_B, γ, and U_T, which are utilized by the **IDsh** instruction described further. The fourth row yields K, the product of the mobility μ times the oxide capacitance per unit area C'_{ox} derived from the oxide thickness t_{ox}. The default value of μ is 500 cm^2/Vs for N-type and 190 cm^2/Vs for P-type transistors (open the **pMat** file to change these). The fifth row represents the gate oxide capacitance per unit area C'_{ox}. Every item can be accessed separately by means of its row index. For instance, p(3) reads U_T or kT/q.

Consider a N-channel transistor with a doping concentration equal to 10^{17} at.cm^{-3} and an oxide thickness of 5 nm. T is equal to 300 °K:

$$p = p\text{Mat}(300, 1e17, 5) \quad (A2.1)$$

pMat outputs one column, the interpretation of which is:

$$
\begin{array}{lll}
p(1) = 0.4078 & V & \Leftarrow \quad \Phi_B \\
p(2) = 0.2646 & V^\wedge(0.5) & \Leftarrow \quad \gamma \\
p(3) = 0.0259 & V & \Leftarrow \quad U_T \\
p(4) = 3,45e - 4 & A/V^\wedge 2 & \Leftarrow \quad K = \mu\, C'_{ox} \\
p(5) = 6,90e - 7 & F/cm^\wedge 2 & \Leftarrow \quad C'_{ox}
\end{array} \quad (A2.2)
$$

The parameters are updated automatically when the temperature changes owing to appropriate expressions stored in the **pMat** file. Consider for instance three temperatures 250, 300 and 350°K (MATLAB A13.m):

$$p = p\text{Mat}(250 : 50 : 350, 1e17, 5); \quad (A2.3)$$

The output consists now of a 5 rows and 3 columns matrix. Each column corresponds to a temperature, 250 first, etc (γ and C'_{ox} are constants of course):

$$
p = \begin{array}{llll}
0.4832 & 0.4462 & 0.4078 & \\
0.2646 & 0.2646 & 0.2646 & \\
0.0173 & 0.0216 & 0.0259 & \\
0.7762 & 0.4968 & 0.3450 & *e\text{-}3 \\
0.6900 & 0.6900 & 0.6900 & *e\text{-}6
\end{array} \quad (A2.4)
$$

A2.1.2 The surfpot(p,V,VG) Function

The **surfpot** function calculates the surface potential by solving the non-linear implicit function listed under Eq. 2.20. The input data are the *Technology vector* p, the non-equilibrium voltage V and the gate voltage V_G. These may be scalars and/or equal length column-vectors.

Consider the same example as above with T equal to 300°K. The gate-to-substrate voltage is constant and equal to 2 V while the non-equilibrium voltage V varies from 0 to 2 V. Two lines suffice in order to evaluate the surface potential, the *Technology vector* given by Eq. A2.1 and the **surfpot** function:

$$psiS = \text{surfpot}(p, linspace(0, 2, 100)', 2); \quad (A2.5)$$

The resulting surface potential is shown in Fig. 3.1. Knowing the surface potential, we can evaluate the threshold voltage V_T given by Eq. 3.6. All what is needed is to add the line below where $p(2)$ stands for γ.

$$V_T = p(2)^* sqrt(psiS) + psiS; \quad (A2.6)$$

A2.1.3 The IDsh(p,VS,VD,VG) *Function*

The ***IDsh*** function evaluates the drain current of 'unary' transistors according to the C.S.M model ('unary' means that the W over L ratio is equal to one). The input data consist of the ***Technology vector*** p and the terminal voltages with respect to the substrate: V_S, V_D and V_G. These may be scalars, equal length vectors or combinations. The function makes use of the MATLAB *polyval* instruction:

$$\frac{I_D}{\beta} = \text{polyval}\left(P, \sqrt{\psi_{SD}}\right) - \text{polyval}\left(P, \sqrt{\psi_{SS}}\right) \quad (A2.7)$$

The surface potentials ψ_{SD} and ψ_{SS} are derived from the **surfpot** function, V being equal to V_D and V_S. P is a row vector consisting of the coefficients ranked from highest to lowest order of the polynomial listed under Eq. 2.19:

$$P = \left[-\frac{1}{2} - \frac{2}{3}\gamma\left(V_G + U_T\right) \quad \gamma U_T \quad 0 \right] \quad (A2.8)$$

It is recommended to add a realistic flat band voltage V_{FB} to V_G to take into consideration interface charges and the gate work function. V_{FB} is a separate variable that makes the gate voltage look more realistic. It does not reside in the ***Technology vector*** and is chosen freely. The flat-band voltage of N-channel transistors lies generally in the range 0.6–0.9 V. It depends on the physical treatments the transistor has been subjected to during fabrication, such as oxidation temperature, ion implantation, etc.

An example illustrating the use of the **IDsh** function is given in Annex 3.

A2.2 Compact Model Files

The files hereafter relate to the compact model of Chapters 4 and 5.

A2.2.1 The Identif 3(Nb,tox,VFB,T) *Function*

The **Identif3** function bridges the C.S.M. to the E.K.V. compact model. The function extracts n, V_{To} and I_{Suo} from C.S.M. drain currents. The input data are the substrate impurity concentration, oxide thickness, flat-band voltage plus the temperature. The parameter extraction is done by means of the algorithm described under Section 4.5.1.

A2.2.2 The **invq(z)** *Function*

The **invq** function inverts Eq. 4.2.3d:

$$V_P - V = U_T \left(2(q-1) + \log(q)\right) \quad \text{(A2.9)}$$

and computes the normalized mobile charge density q

$$q = invq\left(\frac{V_P - V}{U_T}\right) \quad \text{(A2.10)}$$

The pinch-off voltage V_P and non-equilibrium or channel voltage V may be scalars, equal size vectors or matrices.

A2.2.3 The **ComS(VGS,VDS,VS,lg)** *Function*

The function **ComS** evaluates the drain current I_D and the output conductance g_d versus V_{DS} of the variable parameters compact model. The gate-to-source voltage V_{GS} must be a scalar, the drain-to-source voltage V_{DS} a row vector (or a scalar) and the source voltage a scalar. Both, V_{GS} and V_{DS}, can take any value between 0 and 1.2 V whereas V_S should be one of the nine equally spaced source-to-substrate voltages comprised between 0 and 0.8 V.

The function evaluates n, V_{To}, I_{Suo} and the *Theta* function considering for the drain voltage two V_{DS} vectors separated by ± 1 mV. The output conductance g_d is derived from the *diff* of the drain current vectors divided by the 2 mV difference separating the drain voltages. The drain current I_D is the mean of the two drain currents. The output of the **ComS** function consists of a two columns matrix y, the first column represents the drain current I_D, the second the output conductance g_d.

A2.3 Other Functions

A2.3.1 The **jctCap(L,W,R,V)** *Function*

The **JctCap** function evaluates junction capacitances knowing the gate length $L(\mu m)$ and the gate width $W(\mu m)$ of N- and P-channel transistors (see Section 6.2.2). L and W may be scalars, vectors or matrices. The transistors are partitioned automatically in sub-transistors with identical widths comprised between maximal and minimal tolerated values fixed by R and $R/2$ (μm). Partitioning takes place as soon as W gets larger than R. The fourth variable V takes care of the reverse voltage applied to the junction. V is defined with respect to

A2.3 Other Functions

the substrate for N-channel transistors and V_{DD} for P-channel. All capacitances are multiplied by the factor:

$$(1 + V/.5).^\wedge(-0.5)$$

Every capacitance combines a vertical junction capacitance CJ, two peripheral capacitances CJsw (outside periphery) and CJswg (inside periphery – the side capacitance between the junction and the channel) as illustrated in Fig. 6.9. The unit-capacitances are respectively equal to:

1e-15 F/μm^2 for CJ
1e-16 F/μm for CJsw
3e-16 F/μm for CJswg

The **JctCap** function outputs a 5D array y(:,:, 1 to 5) consisting of matrices having the same dimensions as L and W (the sizes of L and W determine the number of colons). These represent:

y(:,:,1) the drain junction cap. CJD
y(:,:,2) the source junction cap. CJS
y(:,:,3) the number of sub-transistors
y(:,:,4) the width of each sub-transistor
y(:,:,5) the total area of the transistor

Source capacitances are always larger than drain capacitances since the first surround the second.

A2.3.2 The Gss(x,H) Function

The **Gss** function calculates the Gaussian distribution of data listed in the column vector x. H is an optional variable representing the mean of x. The Gaussian distribution encompasses the 20 bins histogram of x (opening the **Gss** file allows changing the number of bins called M). The file outputs a graph representing the histogram, Gaussian distribution and lists the 3-sigma of the data in the command window.

Annex 3
Temperature and Mismatch, from C.S.M. to E.K.V.

The influence of temperature and mismatch on the drain current and g_m/I_D of the Charge Sheet Model is examined hereafter. It is extended to the E.K.V. model.

A3.1 The Influence of the Temperature on the Drain Current (MATLAB A31.m)

The influence of the temperature on I_D can be illustrated by means of the **IDsh** function of the MATLAB toolbox. The file below shows an example considering a grounded source transistor undergoing a temperature change from 250 to 350 K. The doping concentration of the substrate is supposed to be equal to 10^{17} at.cm^{-3}, the oxide thickness equal to 5 nm and the flat band voltage equal to 0.8 V. The drain voltage is large enough to keep the transistor saturated while the gate voltage varies from 0 to 2 V.

After inputting technological and electrical data, the **pMAT** function is called in order to set up the Technology Matrix required by the **IDsh** function.

```
% influence of T on ID(VG)
clear
clf
% technological data -----------------------
T = 250: 50: 350; % row vector
N = 1e17;
tox = 5;
VFB = 0.8;
% electrical data --------------------------
VS = 0;
VD = 2;
VG = linspace(0,2,50)'; % column vector.
% compute ---------------------------------
p = pMat(T,N,tox);
for k = 1:length(T),
```

```
ID(:,k) = IDsh(p(:,k),VS,VD,VG + VFB);
end
% plot ------------------------------------
semilogy(VG,ID);
xlabel('V_G (V)'); ylabel('I_D (A)'); grid                    (A3.1)
```

A number of well-known effects are illustrated in Fig. A3.1. When the temperature increases, the drain current grows rapidly in weak inversion while the opposite holds true in strong inversion. Conflicting effects explain the antagonist trends. The influence of the rising temperature on the factor A of Eq. 2.31 explains the increase in weak inversion. Mobility degradation explains the decrease in strong inversion. The first overrules the second in weak inversion while the opposite holds true in strong inversion. Around 0.8 and 1V, the two cancel out.

A3.2 The Influence of the Temperature on gm/ID (Matlab A32.m)

The evaluation the influence the temperature has on g_m/I_D is straightforward since the ratio boils down to the slope of the curves plotted in Fig. A3.1. The result is

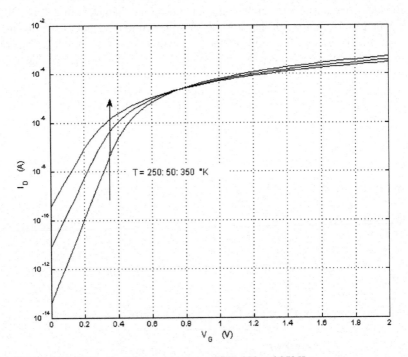

Fig. A3.1 C.S.M. drain current for temperatures of 250, 300 and 350 K

A3.2 The Influence of the Temperature on gm/ID

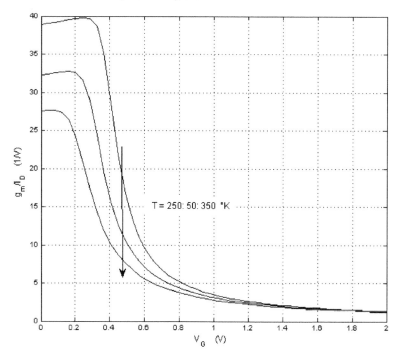

Fig. A3.2 g_m/I_D versus temperature of the transistor considered in the previous figure

Table A3.1 Comparison of temperature sensitivities of g_m/I_D's

T(K)	$n_{w.i.}$ (Eq. 2.38)	U_T 0.026.T/300	$1/nU_T(\mathrm{V}^{-1})$	$\max(g_m/I_D)$ (C.S.M.)
250	1.1628	0.0217	39.69	39.79
300	1.1749	0.0260	32.74	32.75
350	1.1892	0.0303	27.72	27.62

displayed in Fig. A3.2. The lessening of the subthreshold slope in weak inversion has a strong impact on the maximum g_m/I_D.

Table A3.1 compares the maximum of g_m/I_D predicted by the C.S.M. (most right column) to $1/nU_T$. The first is derived from the maximum of the derivative of $\log(I_D)$ whereas the second takes advantage of the analytic expression of the slope factor given by Eq. 2.38. The table shows that the latter is clearly a good approximation of the C.S.M. slope factor.

A3.3 Temperature Dependence of E.K.V Parameters (MATLAB A33.m)

We showed in Chapter 4 that the basic E.K.V model is an approximation of the C.S.M. The acquisition method enabling to extract E.K.V parameters from C.S.M drain currents described in Section 4.5 offers the possibility consequently to assess the impact of the temperature of n, V_{To} and I_{Suo}. The plots of Fig. A3.3 show the influence of the temperature of the slope factor n, the threshold voltage V_{To} and the unary specific current I_{Su} when the temperature goes from 250 to 350 K. The threshold voltage, which is equal to 0.3984 V at 300 K, drops by 1.31 mV/°C, the slope factor, equal to 1.1267, increases by 8.2×10^{-5} per°C, and the unary specific current, equal to 4.44×10^{-7} A, decreases by 62.3 pA per°C.

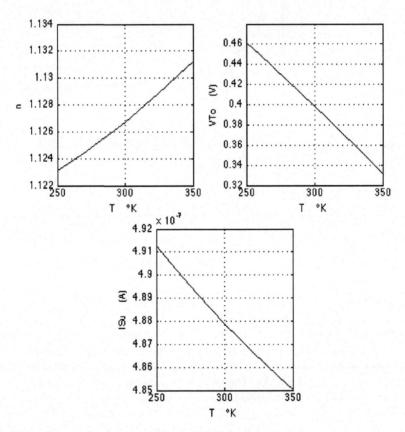

Fig. A3.3 influence of the temperature on the E.K.V parameters

A3.4 The Impact of Technological Mismatches on the Drain Current (Matlab A34.m)

The impact of substrate doping and oxide thickness mismatches on the drain current can be assessed easily with the C.S.M model. We consider the same transistor as above with T equal to 300 K and suppose that the doping concentration N and the oxide thickness *tox* obey Gaussian distributions, with sigmas respectively equal to 2.0% and 0.5%. We consider two constant gate voltages, one in weak and one in strong inversion. The two histograms of Fig. A3.4 give an idea of the spread of the drain current caused by the mismatches. Left, the gate voltage is equal to 0.2 V, right it is equal to 0.6 V. The mean unary drain currents are respectively 5.56 nA and 8.87 µA. The high mismatch sensitivity of MOS transistors in weak inversion is corroborated by a large spread.

```
% influence of N and tox mismatch on ID(VG)
clear
clf
% technological data ----------------
T = 300;
z = 1000; % number of samples
N = 1e17*(1 +.02*randn(1,z));
tox = 5*(1 +.005*randn(1,z));
VFB = 0.8;
% electrical data
VS = 0;
VD = 2;
VG = 0.2;
% compute
for k = 1:z,
p = pMat(T,N(1,k),tox(1,k));
ID(:,k) = IDsh(p,VS,VD,VG + VFB);
end
% plot -----------------------------
M = mean(ID);
[n,x] = hist(ID(1,:),10);
bar(x/M,n)
h = findobj(gca,'Type','patch');
set(h,'FaceColor','w','EdgeColor','k')
axis([0 1.5 0 300]);
xlabel('I_D/mean(I_D)');
ylabel('histogram 1000 samples');
text(.3,200,'V_G = 0.2 V')
```

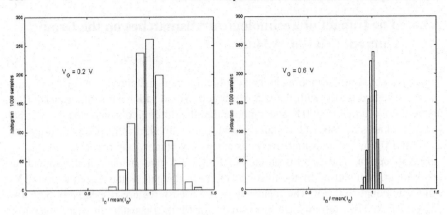

Fig. A3.4 Comparative histograms of relative drain currents spreads *left*, V_G is equal to 0.2 V (weak inversion), *right*, V_G is equal to 0.6 V (stong inversion)

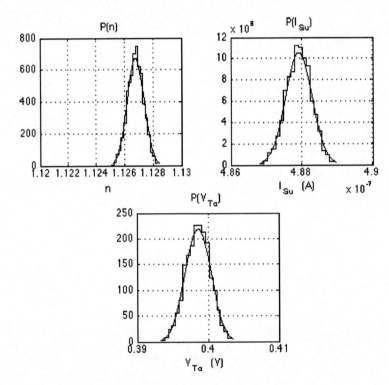

Fig. A3.5 Probability densities of E.K.V. model parameters

A3.5 Mismatch and E.K.V Parameters (MATLAB A35.m)

Since the C.S.M. offers the possibility to evaluate the impact of mismatches on drain currents, we can also evaluate their impact on the parameters of the equivalent E.K.V. model. We consider a Gaussian mismatch of the substrate impurity concentration centered around 10^{17} at.cm^{-3}. The sigma is equal to 1%. The oxide thickness and flat band voltage are constant and the same as in the previous example. The probability densities of n, V_{To} and I_{Suo} are displayed in Fig. A3.5.

The impact of mismatches on the parameters is illustrated by the three-sigma deviations listed below:

$3\sigma(n) = 0.0018\%$
$3\sigma(V_{To}) = 5.6\,\text{mV}$
$3\sigma(I_{Su}) = 1.16\,\text{nA}$

Annex 4
E.K.V. Intrinsic Capacitance Model

The intrinsic gate-to-source and gate-to-drain capacitances of the E.K.V model are compared to their 'semi-empirical counterparts in this annex. We consider a grounded source N-channel transistor and sweep the gate and drain voltages from 0 to 1.2 V. The 'semi-empirical' capacitances are extracted from the global variables **CGSn** and **CGDn** (Courtesy of IMEC). We make use of the expressions below for the model, where C_{ox} stands for the oxide capacitance fixed by the width and the length of the transistor (Section 5.3.1 of Enz and Vittoz 2006):

$$C_{gsi} = C_{ox} \frac{q_F}{3} \cdot \frac{2q_F + 4q_R + 3}{(q_F + q_R + 1)^2} \tag{A4.1}$$

$$C_{gdi} = C_{ox} \frac{q_R}{3} \cdot \frac{2q_R + 4q_F + 3}{(q_F + q_R + 1)^2} \tag{A4.2}$$

To evaluate q_F and q_R versus the gate and drain voltages, the E.K.V. parameters are extracted first from 'semi-empirical' drain currents by means of the acquisition algorithm presented in Chapter 5.

The 'semi-empirical' capacitances include overlap capacitances that are ignored by the E.K.V model. To separate extrinsic from intrinsic 'semi-empirical' capacitances, we evaluate the 'semi-empirical' capacitances under bias conditions minimizing the contribution of the intrinsic capacitances. For instance, we get rid of the inversion layer by zeroing the gate-to-source voltage to evaluate the gate-to-source overlap capacitance C_{gsov}. The fact that the overlap capacitances per μm gate width listed in Table A4.1 are not affected by gate lengths changes while the gate capacitances per μm do, supports the idea.

Figures A4.1 and A4.2 compare 'semi-empirical' (left) to model intrinsic capacitances (right) considering two gate lengths: 500 and 100 nm. To make a fair comparison, we add the gate-to-source overlap capacitance derived from the 'experimental' data to the intrinsic capacitances of the model and adjust the vertical scale to get the same maximum capacitance. It is clear that the E.K.V. intrinsic gate-to-source capacitance is not a bad representation, except when the transistor is not saturated.

Caution is needed however as far as the overlap capacitances. These depend not only on extrinsic contributions but also on the underlying junction-to-substrate

Table A4.1 Extrinsic and intrinsic gate-to-source capacitances (exper. data)

L (μm)	C_{gsov} (fF/μm)	C_{gsi} (fF/μm)
0.100	0.363	0.413
0.110	0.413	0.469
0.120	0.413	0.574
0.130	0.413	0.683
0.140	0.412	0.792
0.160	0.412	1.010
0.500	0.408	4.806
1.00	0.408	10.258
4.00	0.419	42.189

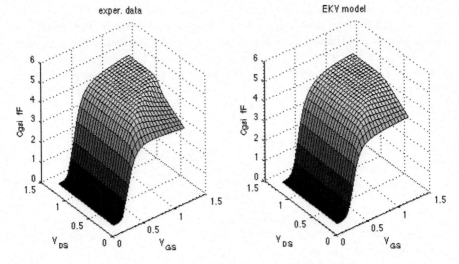

Fig. A4.1 The gate-to-source capacitance of the 500 nm gate length transistors

voltage (see Section 10.3 of Enz and Vittoz 2006). The phenomenon is clearly visible in Figs. A4.3 and A4.4, which displays gate-to-drain overlap capacitances C_{gdov} according to the method above.

When the transistor is saturated, the gate-to-drain capacitance is far from being constant, especially when the gate length effects are not visible on the gate-to-source. The gate-to-drain capacitances predicted by the model is a poor representations of C_{DS} when the transistor is saturated contrarily to C_{GS}.

4 E.K.V. Intrinsic Capacitance Model

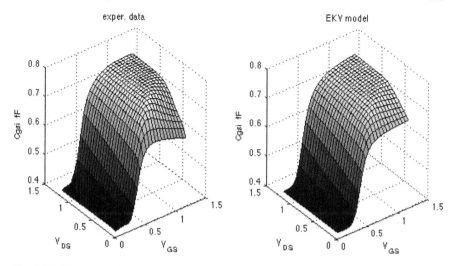

Fig. A4.2 The gate-to-source capacitance of the 100 nm gate length transistors

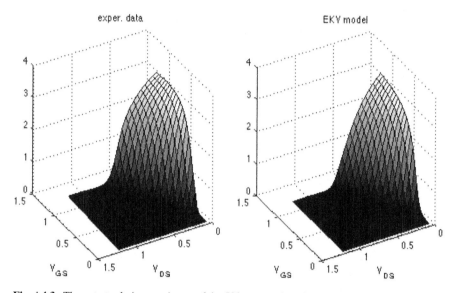

Fig. A4.3 The gate-to-drain capacitance of the 500 nm gate length transistor

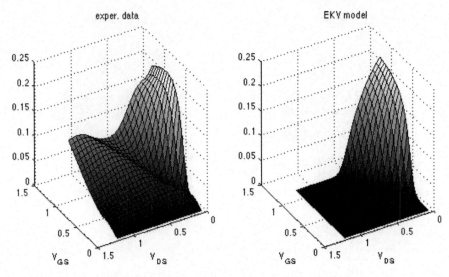

Fig. A4.4 The drain-to-source capacitance of the 100 nm, gate length transistor

Bibliography

Banu M, Tsividis Y (Oct 1984) Detailed analysis of nonidealities in MOS fully integrated active RC filters based on balanced networks. Electron Circuits Syst, IEE Proc G 131(5):190–196

Binkley DM (2007) Tradeoffs and optimization in analog CMOS design. Wiley, Chichester, England, ISBN 978-0-470-03136-0

Binkley DM et al (Feb 2003) A CAD methodology for optimizing transistor current and sizing in analog CMOS design. IEEE Trans Comput Aided Des Integr Circuits Syst 22(2):225–237

Brews JR (1978) A charge sheet model for the MOSFET. Solid State Electron 21:345–355

Brun Y, Lee K, Shur M (Jan 1990) Unified charge controlled model and subthreshold current in heterostructure field effect transistors. IEEE Electron Device Lett 11(1):50–53

Bucher M (1999) Analytical MOS transistor modelling for analog circuit simulation. Ph.D. thesis no. 2114, Swiss Federal Institute of Technology, Lausanne (EPFL)

Cand M, Lardy JL, Demoulin E, Senn P (1986) Conception des circuits integers. Annexe 1, Eyrolles, Paris, pp 163–169

Chatelain JD (1979) Traité d'Electricité, vol VII, Dispositifs à Semiconducteurs, EPFL

Coltinho RM, Spiller LH, Schneider MC, Montoro CG-M (2001) Metodologia simplificada de extraçao de Parâmetros para o modelo A.C.M. do transistor MOS. VII workshop de Iberchip – IWS'2001

Cunha AIA, Scheider MC, Galup-Montoro C (Oct 1998) An MOS transistor model for analog circuit design. IEEE JSCC 33(10):1510–1519

Enz CC (2008) A short story of the EKV MOS transistor model. IEEE Solid State Circuits News 13(3):24–30. www.ieee.org/SSCS-news

Enz CC, Vittoz EA (2006) Charge-based MOS Transistor Modeling. *The EKV model for low-power RF IC design.* Wiley, Chichester

Girardi A, Bampi S (2006) Power constrained design optimization of analog circuits based on physical g_m/I_D characteristics. 19th IEEE Symposium on Integrated Circuits and Systems, SBCCI 2006, pp 89–93

Girardi A, Cortes FP, Bampi S (2006) A tool for automatic design of analog circuits based on g_m/I_D methodology. IEEE ISCAS 2006

Grabinski W, Nauwelaers B, Schreurs D (2006) Transistor level modelling for analog/RF IC design. Springer, The Netherlands

Jespers PGA, Jusseret C, Leduc Y (June 1977) A fast sample and hold charge-sensing circuit for photodiode arrays. IEEE JSSC SC-12(3):232–237

Laker KR, Sansen WMC (1994) Design of analog integrated circuits and systems, McGraw-Hill series in Electrical and Computer Engineering

Miller JM (1920) Dependence of the input impedance of a three-electrode vacuum tube upon the load in the plate circuit. Scientific Papers of the Bureau of Standards, vol 15, no. 351, pp 367–385

Muller RS, Kamins ThI (1977) Device electronics for integrated circuits, 2nd edn. Wiley, New York, p 36

Oguey H, Cserveny S (1982) Sonderdruck aus dem Bulletin des SEV/VSE, Bd. 73, 1982, pp 113–119

PSP (2006) http://pspmodel.asu.edu http://www.nxp.com/Philips_Models/mos_models/psp/

Silveira F, Flandre D, Jespers P (Sept 1996) A g_m/I_D based methodology for the design of CMOS analog circuits and its application to the synthesis of a silcon-on-isulator micropower OTA. IEEE J Solid State Circuits 31(9):1314–1319

Tsividis Y (1999) Operation and modelling of the MOS transistor, EE series. Mc-Graw Hill, New York

Van de Wiele (Dec 1979) A long channel MODFET model. Solid State Electron 22(12):991–997

Vittoz E, Fellrath J (June 1977) Small signal model of MOS transistors in Weak Inversion Operation, JSSC IEEE 12(3):232–237

Wallinga H, Bult K (June 1989) Design and analysis of CMOS analog signal processing circuits by means of a graphical MOST model. JSSC IEEE J 24(3):672–680

Index

A
A.C.M. model, 41

B
body effect, 79
Boltzmann statistics, 13, 25, 42

C
cascoded Intrinsic Gain Stage
  gain evaluation, 117
cascoded Intrinsic Gain Stage
  frequency response, 118
  poles and zeros, 118
  sizing, 115
Channel length modulation (C.L.M.), 78, 80
Charge Sheet Model (C.S.M.), 11
  common-gate configuration, 23
  drain current equation, 13
  drain current versus drain voltage, 15
  drain current versus gate voltage, 17
  g_m/I_D, 20
  weak inversion approximation, 18
common-gate configuration
  compact model drain current and g_{ms}/I_D, 113
compact model for real transistors, 68
  equations, 70
  g_d/I_D, 88
  g_m/I_D, 85
  $I_D(V_{DS})$, 82
  mobility degradation polynomial, 73
  parameter acquisition, 70
  parameter dependence on bias conditions, 78
  parameter dependence on the gate length, 76
cut-off angular frequency, *see* cut-off frequency
cut-off frequency, 3

D
diffusion current, 12, 18
drain induced barrier lowering (D.I.B.L.), 68, 79, 89
  impact on the pinch-off voltage, 83
drift current, 12, 18

E
E.K.V. model, 41
  drain current, 45, 48
  equations, 46
  g_m/I_D, 54
  g_{ms}/I_D, 57
  mobility degradation, 59
  parameter acquisition, 50
  weak and strong inversion approximations, 50
Early voltage, 2, 89
extrinsic capacitances, 99
  transistor partitioning, 101

G
gain-bandwidth product, 3
gate voltage overdrive (G.V.O.), 6
global variables, 143
  compact model parameters, 144
  example calculate $I_D(V_{GS})$ from compact model, 147
  example extract g_m/I_D from semi-empirical data, 144
  semi-empirical, 143
g_m/I_D sizing methodology, 7
gradual channel approximation, 11, 41, 67

graphical construction, 27
    CMOS transmission gates, 35
    compact model, 47
    implementation of linear resistors, 36
    small signal transconductances, 34
    source bootstrapping, 37
    the CMOS inverter, 33
    the MOS diode, 32
    the MOS source follower, 32

I

intrinsic capacitances (E.K.V. model), 163
Intrinsic Gain Stage (I.G.S.), 1
    equivalent circuit, 1
    frequency response, 1
    gain evaluation with var. param. compact model, 106, 107
    simplified sizing procedure making use of the var. param. compact model, 110
    sizing in moderate inversion, 5
    sizing in strong inversion, 4
    sizing in weak inversion, 4
    sizing the cascoded I.G.S., 115
    sizing with E.K.V model, 55
    sizing with E.K.V. model and mobility degradation, 65
    sizing with semi-empirical data (constant output capacitance), 95
    sizing with semi-emprical data (with output junction capacitance), 103
    sizing with variable param. compact model, 104
    transfer function, 108

J

junction capacitances
    vertical and side-wall capacitances, 101

M

MATLAB
    IDsh function, 15
    Indentif3.m function, 50
    pMat function, 15
    surfpot function, 15
MATLAB toolbox, 149
Miller Op. Amp., 121
    analysis, 122
    current mirror, 126
    frequency response, 124
    phase margin, 129
    pole splitting, 123
    poles and zeros, 127
    sizing a high-frequency low-power Miller Op. Amp., 140
    sizing a low-voltage Miller Op. Amp., 130
    sizing methodology, 129
    transfer function, 127
mismatch, 155
mobility coefficient, 12
mobility degradation, 80, 83
    critical field, 60
    first order approximation, 59
    impact of mobility degradation on the drain current, 60
    impact of mobility degradation on g_m/I_D, 64
    impact on the specific current, 80
    longitudinal and vertical electrical fields, 80
MOS
    quadratic model, 4
    weak inversion model, 4

N

normalized drain current, 45
    forward, 46
    reverse, 46
normalized mobile charge density, 42

P

pinch-off voltage, 27, 38, 43, 44, 62, 83

Q

quasi-stationarity, 98

R

reverse short channel effect, 67, 78, 79
roll-off, 67, 78, 79

S

semi-empirical g_m/I_D, g_d/I_D and gain dependence on bias conditions, 93
short channel effects, 67
sizing-space dimensions, 121
slew-rate, 7, 111, 142
slope factor, 41
specific current, 45
    unary specific current, 50
specifications and attributes, 121
subthreshold slope, 18, 22
surface potential, 12

T

temperature, 155
threshold voltage, 24
    of E.K.V. model, 45
    with respect to the source, 26
    with respect to the substrate, 26, 28, 30, 31
transistor partitioning, 101
transition angular frequency, 3
transition frequency, 3

Printed by Printforce, the Netherlands